I0794574

MIRAGE, THE COMBAT LOG

MIRAGE, THE COMBAT LOG

by Salvador Mafé Huertas

Schiffer Military/Aviation History
Atglen, PA

ACKNOWLEDGEMENTS

To make a book about such a subject as the Mirage and its derivatives has involved many persons, and in the following lines, I will try to remember all which helped in one way or another, without them this book could not have been possible.

Javier Matute, Javier Valero, Fernando Ribes, Manuel Perales, Jose Sanz, Miguel Sales and Federico Wichi of the famed Alambique (113) Sqn; also all the members of the Mirage equipped units in the Spanish Air Force, 111, 141, 142 and 462 Squadrons as well as Major Francisco Javier Fernandez of 141 Squadron, with more than 2000 hours on type who showed me the paces of flying the Mirage F1 and Colonel Yago F Bobadilla, "my" first F1 pilot way back in 1981, and today in charge of the Spanish F1 Upgrade Programme; Jesus Romero Briasco, C.O. of 111 Sqn., Spanish AF in 1986-87; also to Javier Muñoz Castresana, the last C.O. of 111 Sqn. while still equipped with Mirage IIIs. Spanish AF Public Relations Office; Armée de l'Air Public Information Officer; Major Laverne Mills, Senior Officer Media Liason SAAF; Brigadier Johan Rankin, SAAF; Israel Defence Force Spokesman; Israel Aircraft Industries; Atlas Aviation Ltd; ENAER; Shlomo Aloni; Peter Steineman; Roger Soupart; Inter-Service Public Relations (Pakistan); Historical Branch of the Argentine Air Force, plus to many of the pilots who fought during the Falklands War, as well as their adversary Sea Harrier jocks, on the other hand, a special thank you is deserved for the unknown and unidentified former Iraqi Air Force F1 pilot, which for obvious reasons has to remain anonymous; Although accounted before, a note of special gratitude are to go to 111 Squadron pilots, Joaquín Díaz, Carlos Sebastia, Rafa Nuñez, Felix Sahagun, Jorge Albalat, Jesus Labajo, Javier Bas, Agustin Serrano, Sergio Riego, Jesus Elices, Jesus Ruiz, Cesar Lardies, Paco Almenta, Rafa Suevos and Rafa Mico, as well as the ground support team, including Felipe Ramajo, Manuel Rosa, Juan Campos, Miguel Angel de Andres and all the others; finally former 111 Squadron pilots Carlos Pinilla and Chemi Quintana, were of much help also. Finally, I don't want to leave out Captain Jorge Albalat who flew me in the Mirage F1DDA of 111 Sqn. in a very interesting operational sortie.

Dedicated to my daughters Mercedes and Cristina,
and to my friend Pitu II, who took-off on his last flight on March 24, 1996.

Book Design by Ian Robertson.

Copyright © 1996 by Salvador Mafé Huertas
Library of Congress Catalog Number: 96-68258

All rights reserved. No part of this work may be reproduced or used in any forms or by any means – graphic, electronic or mechanical, including photocopying or information storage and retrieval systems – without written permission from the copyright holder.

Printed in China.
ISBN: 0-7643-0168-3

We are interested in hearing from authors with book ideas on related topics.

Published by Schiffer Publishing Ltd.
77 Lower Valley Road
Atglen, PA 19310
Phone: (610) 593-1777
FAX: (610) 593-2002
Please write for a free catalog.
This book may be purchased from the publisher.
Please include $2.95 postage.
Try your bookstore first.

Contents

Introduction

This is the first book ever produced about all the Mirage variants and their record in combat operations, from the Mid-East to the South Atlantic, from the Gulf, to the steaming jungles between Ecuador and Peru. Technical and development details of the several Mirage families have been kept to a bare minimum, and no attempt has been made to describe experimental variants, nor the Mirage IV strategic bomber, which escapes the purpose of this book narrating the combat career of the fighter versions. The contrary that is done about the operational details of these extraordinary creations from the Dassault Design Bureau.

As the reader will notice, in chapters two and three, A LEGEND IS BORN, THE ISRAELI MIRAGES and THE SIX-DAY WAR, I have included not only the chronology about the three Mirage squadrons, but also about the other combat squadrons of the Israeli Air Force, in first place as this will provide a better general picture about this valiant air arm, in second place, and concerning chapter three, I did it because the actions of the Chel Ha'Avir during the June 1967 War established a doctrine about well coordinated strikes against an enemy superior in number, and from then on it was a lesson to be learned in Air Staffs around the world. Many personal narratives are provided by Mirage pilots, Israeli, Argentine, South African as well as, and this is most interesting, an exiled Iraqi Air Force pilot who used to fly Mirage F1s. All in all, this is a book about three generations of fighters and one bomber generation, as well as their derivates produced in other countries. But mainly this is a book about one of the western world's jets that has been in more wars and conflicts than any other, and that's what the reader will find inside, combat flying to the limits, with the plus of a careful selection of photographs, most of them published here for the first time. Since 1961 more than 2,967 Mirage fighters of three generations have been produced in France and abroad, with and without license, only the Mirage III/5/50 family accumulated more than three million flight hours from July 1961 to July 1994.

Now dear reader, strap-on to the Martin Baker ejection seat of one of these French beauties and fly into the wild blue yonder.

Salvador Mafe Huertas
Valencia
May 1996

IN MEMORIAM

This book is dedicated to the memory of my friend Christian de Haya Beyrer, former pilot of Mirage F1s with 111 Squadron, Spanish Air Force, tragically killed while flying a CASA C-101EB Aviojet trainer on 13 March 1995.

1

The Mirage Saga

This chapter describes, in photographic form, all the Mirage variants — Mirage III, 5, F1, and 2000 — with detailed captions describing subvariants and users. Also, in three boxes, see below — descriptions of the specifications of the Mirage IIIE, F1C, and Mirage 2000C.

SPECIFICATIONS - MIRAGE IIIE

Wing: span 26 ft. 11.6 in.; aspect ratio 1.94; area 376.75 sq. ft.
Fuselage and tail: length 49 ft. 3.5 in.; height 14.9 in.; wheel track 10 ft. 4 in.; wheel base 15 ft. 11.75 in.
Weights: empty 15,542 lb.; normal take-off 21,164 lb.; maximum take-off 30,203 lb.
Fuel and load: internal fuel 631.4 U.S. gal. increased by 145 U.S. gal. if the rocket pack is not fitted; external fuel up to two 449, 343, 291 or 165 U.S. gal. drop tanks, or two 66 U.S. gal. non-jettisonable supersonic tanks, or two 66 U.S. gal. JL-100 combined drop tanks/rocket launchers; maximum ordnance 8,500 lb.
Powerplant: one SNECMA Atar 9C-3 rated at 9,436 lb. dry and 13,670 with afterburning, and provision for one SEPR 844 rocket booster rated at 3,300 lb.
Speed: Maximum level speed clean at 40,000 ft. 1,270 knots; cruising speed at 36,000 ft. 516 knots.
Range: ferry range 2,152 n.m. with three drop tanks; combat radius 647 miles on a hi-hi-hi mission with two Magic/Sidewinder AAMs.
Performances: maximum rate of climb at sea level more than 16,405 ft. per minute; climb to 36,000 ft. in 3 minutes; service ceiling 57,000 ft. or 75,500 with rocket pack; take-off run between 2,300 and 5,300 ft. depending on its weight; landing run 2.300 feet with brake chute; g limits +5 in a sustained turn at Mach 0.9 at 16,500 ft.

Weapons: Cannon armament of two 30-mm DEFA 552s with 125 rounds per gun. MATRA R.530 AAM on centerline pylon, Hughes AIM-26 Falcon on Swiss Mirages. Two wing pylons for infrared guided missiles, either AIM-9 Sidewinders or MATRA R.550 Magics. Mirage IIIE multirole capability introduced a maximum of five pylons — seven on most upgraded aircraft — with a maximum weapon load of 8,500 lb., including the majority of free fall bombs and rocket pods. While the missiles used were the Aerospatiale AS30 and MATRA AS37 Martel, the French Mirage IIIE could carry the AN52 tactical nuclear bomb.

The small Dassault MD550 Mystère Delta was the first in a proud line, which includes the Mirage III, 5, 50, F1 and even the 2000. It was immediately recognizable by its compact size, two engines and broad chord tail. (Dassault).

Dassault built ten pre-series Mirage IIIAs used for all the development work, including airframe, weapons, engine and radar, which led to the first operational model, the Mirage IIIC.
Illustrated is Mirage IIIA number 07 in clean configuration, during a test sortie from Istres in the south of France. (Dassault)

Developed from the Mirage IIIC was the twin-stick Mirage IIIB conversion trainer, which carried slightly less fuel but also lacked the Cyrano Ibis fire control radar. This aircraft, serial 224 in the early natural scheme belonged to Escadro de Chasse et de Transformation 2/2 "Côte d'Or" based at Dijon air base, which for many years was the Mirage III Operational Conversion Unit for French and many foreign pilots. (ECT 2/2)

The original Dassault built Australian Mirage IIIO prototype was powered by a Rolls-Royce RA.146 Avon 67 engine with plans to install this engine to the future RAAF Mirage III fleet, and although performance gains were quite evident, finally the Australian Mirages carried the standard Atar 9C engines. This particular aircraft was named "City of Hobart". On December 7, 1964, while assigned to the Air Research and Development Center (ARDU) the machine was written-off in a crash at Avalon, Australia. (Dassault)

The RAAF received a total of 100 single seat Mirage IIIOs (both in the "F" and "A" - all later converted to the "A" subvariant) as well as 16 twin-sticker Mirage IIIDs, serving in all with five squadrons, numbers 3, 75, 76, 77 and 79, plus No 2 Operational conversion Unit, in the period 1964 to 1989. Most Mirages were assembled in Australia by Government Aircraft Factories of Avalon as the primary contractor. The picture shows six Mirage IIIO(A)s from No 75 Squadron, based at RAAF Tindal, northern Australia, indeed the last RAAF Mirage squadron. It's last operational mission was flown on 28 September 1988. (RAAF)

Between September 1967 and March 1969 The Lebanese Air Force received ten Mirage IIIELs and two Mirage BLs, they were scarcely used and spent most of the time stored, even during the Lebanese Civil War. Some sources claim that they, together with their full spare stock and the supply of Matra R530 missiles went to Syria, others claim that they found their way to Pakistan. (Dassault)

In RAAF service the Mirage III ended their days with the colorful ARDU fleet at Avalon, which kept one single seat (A3-2) and 2 two-seaters (A3-101 and 112) in service until late 1989. The picture shows A3-112 in its colorful "Fanta can" scheme. (Meggs)

Egypt's first batch of 54 Mirage 5SDEs (basically Mirage IIIEs) were paid for by Saudi Arabia, and for political reasons wore Royal Saudi Air Force markings until delivered to Egypt, as can be seen in this picture of Mirage 5SDE number 1013 with Saudi markings just before departing for the ferry flight for Egypt. (Cuny)

The Mirage IIIBE (IIID for export) was the two seat version of the Mirage IIIE multirole fighter, although it lacked the Cyrano IIbis radar and Marconi-Elliot doppler navigation radar, it had the two DEFA 522 30mm guns with 125 rounds each. The IIIBE shown here belonged to Escadron de Chasse et the Transformation 2/2 at Dijon, carrying on the left side of its tail the sinister Grim Reaper badge of SPA94 (the squadron's 3rd Escadrille), applied over a black chevron. (Dassault)

The Swiss Air Force and Anti-aircraft Command received a total of 36 Mirage IIIS in the period 1964-1969, 34 of which were assembled under license by the Swiss F+W company, these aircraft were delivered with the unique Hughes Taran 18 fire control radar (in place of the Cyrano) and included the Hughes AIM-4 Falcon radar guided AAM, and Aérospatiale AS 30 ASM among its main weapons, besides Sidewinders and free-fall bombs. These machines equipped - and still equip, although shortly they will be supplanted in the air defense role by the F-18C/D Hornets, Fliegerstaffeln 16 and 17 at Payerne, near Zurich. The 30 surviving machines were modernized in the ISMA (Improved Swiss Mirage Aircraft) including canards, as well as a completely new ECM fit. On the other hand the Swiss received a Mirage IIICS -used for test purposes-, four Mirage IIIBS and two IIIDS two-seaters. The Mirage IIIS illustrated (J-2302) was used for development purposes during the ISMA program, note the mission stencilling on the nose (F+W)

The Swiss also put into service 18 Mirage IIIRS recce fighters, which also were involved in the ISMA program, as can be seen in this picture, and these actually serve with Fliegerstaffeln 3, 4 and 10. (F+W).

The ex-Israeli Mirage 5J -renamed in French service as Mirage 5Fs- entered French service in 1972 and was phased out in 1994. In addition to the 50 in use, another eight were produced and delivered in 1983-84, to replace eight Mirage 5Fs sold to Chile as 5FC, in fact the last of these, number 58, was the last of 465 Mirage III/5s supplied to the French Air Force. Shown here is an aluminum Mirage 5F, and specially painted to celebrate the 20th Anniversary of Escadron de Chasse 3/13 Auvergne, which took place at its base, Colmar, in June 1992. (Willemsen)

Gabon originally ordered seven Mirages, comprised of three Mirage 5G fighters (shown here), two 5DG trainers and two 5RG recce aircraft, although the latter were canceled before delivery. Heavy operational attrition resulted in the delivery of two 5G-2 fighters and two 5DGs, while the two surviving 5Gs were upgraded to 5G-2s. Deliveries of the first batch took place in 1978 and of the second batch in May 1984. (Dassault)

The Mirage 5E2 was the last of the Mirage deltas delivered to Egypt, a total of sixteen aircraft served with an attack regiment, the last being delivered in 1984. These Mirages have revised avionics, incorporating many features of the later Mirage F1, including HUD and laser rangefinder. (Soupart)

Zaire received eight Mirage 5Ms and three Mirage 5DMs delivered in 1975 and equipping 211 Escadrille de Chasse at Kamina, the unit was deployed to Chad in the mid-eighties in that nation's struggle with Libya, flying several combat sorties, one of them being written off (M402) in a landing accident. (Cuny)

The United Arab Emirates received 32 first generation Mirages, comprising 12 Mirage 5AD, three Mirage 5DAD (illustrated), 14 Mirage 5EAD and three Mirage 5RAD, operating from al Dhafra, they formed two units I Shaheen Squadron, equipped with Mirage 5EADs and 5DADs, and II Shaheen Squadron with Mirage 5ADs and 5RADs, the later reequipping in 1989 with Mirage 2000s. (Cuny)

Belgian Mirages have been painted in gaudy schemes for anniversary celebrations, as can be seen on this Mirage 5BA from 8 Escadrille in a colorful white/blue scheme with the number eight painted undersides. (de Bruyn)

Grupo 4 of the Chilean Air Force has received eight Mirage 50FCs, six Mirage 50Fs and three Mirage 50DCs since being recommissioned on 15 November 1980, at Arturo Moreno Benitez Airport, Santiago, moving six years later to Carlos Ibañez air base, near Punta Arenas in the far south, the whole fleet is in the process to be upgraded to Pantera standards, in a configuration and with avionics similar to those of the Kfir C7, and number 514, seen at Carlos Ibañez air base, was the first of them, note Shafrir IV AAMs. (Steinemann)

The Dassault Balzac was a V/STOL Mirage III, only one technology demonstrator was constructed, carrying a battery of eight lift jets for landings and take-offs, the formula did not succeed and the project was abandoned.

The Mirage IIIV, of which two prototypes were built used the same formula as the Balzac but was twice the size and weight, but despite good flying performances, the lifting engine concept was viewed as impractical and the project was abandoned. (Dassault)

The Mirage IIIT was a scaled-up trials aircraft designed to test the TF104 turbofan intended for the Mirage IIIV, tests continued until 1970, but then was stored and ended its days as a gate guardian. (Dassault)

The Mirage IIING prototype, which later became the Mirage 50M demonstrator, had previously served as one of the Mirage 50 prototypes, and originally it was a Mirage IIIR. This upgraded variant had many of the Mirage 2000 avionics, as well as its fly-by-wire system, but Dassault did not find any customer for this new variant. (Dassault)

SPECIFICATIONS - MIRAGE F1C

Wing: span 27 ft. 6.75 in. without tip AAM rails and 30 ft. 6.75 in. with tip mounted Sidewinders; aspect ratio 2.82; area 269.11 sq. ft.
Fuselage and tail: length 50 ft. 2.5 in.; height 14 ft. 9 in.; wheel track 8 ft. 2.5 in.; wheel base 16 ft. 4.75
Weights: empty 16,314 lb.; normal take-off 24,030 lb.; maximum take-off 35,715 lb.
Fuel and load: internal fuel 1,136 U.S. gal; external fuel up to one 581 U.S. gal. and two 298 U.S. gal. drop tanks; maximum ordnance 13,889 lb.
Powerplant: one SNECMA Atar 9K-50 engine rated at 11,023 lb. dry and 15,785 lb. with afterburning.
Speed: maximum level speed clean at 36,000 ft. 1,262 knots.
Range: combat radius 229 n.m. on a hi-lo-hi attack mission with 14 551 lb. bombs or 324 n.m. on a lo-lo-lo attack mission with six 551 bombs and two drop tanks, or 749 n.m. on a hi-lo-hi attack mission with two 552 lb. bombs and three drop tanks; endurance 2 hours 15 minutes on a CAP with two Super 530F-1 AAM and one drop tank.
Performance: maximum rate of climb at sea level 41,930 ft. per minute with afterburning; service ceiling 66,000 ft.; take-off run 1,960 ft.; landing run 2,200 ft. at 18,739 lb.; g limits +7.5
Weapons: two DEFA 553 30-mm cannon with 135 rounds per gun; standard air-to-air configuration of two AIM-9 or Magic missiles on wingtip rails, as well as two Super 530F-1s on wing pylons. Ground attack capability in the nav/attack software, with many types of free fall bombs, including Laser Guided Bombs.

The Dassault Mirage IIIF2 was a scaled-up Mirage III with a shoulder swept wing and a conventional tail, one prototype was built, flown in June 1966, to meet an Armée de l'Air specification for a supersonic strike aircraft, but the project was eventually canceled. (Dassault)

The prototype of the original Mirage F1E, which was powered by a SNECMA M53 turbofan (the same that powered the Mirage 2000), being developed as a private venture export fighter, although the Armée de l'Air was soon persuaded to place an order to give the project the added credibility of having a home customer. Dassault fought hard to obtain orders from the four European nations (Belgium, Denmark, Netherlands and Norway) participating in the "contract of the century" stressing the all-European nature of the project. The competition was won by the F-16 Fighting Falcon and the F1E was no further developed, although its designation was a applied to the export multirole variant. (Dassault)

The second prototype of the Mirage F1 (foreground) flying formation with the two-seat Mirage IIIF2, also known as F2. Eventually only the F1 would see production, with more than 700 examples completed by Dassault. (Dassault)

The Mirage IIIG8 was a swing wing variant, both prototypes sported different configurations, one being a single seater and the second a two seater, both being twin engine machines, although with high potential they were not further developed due to budgetary cuts. (Dassault)

Dassault's assembly line for the Mirage F1, in the foreground are two French F1C-200s, followed by an Iraqi F1EQ, an Spanish F1CE and finally another Iraqi F1EQ. (Dassault)

This air-to-air view shows one of the first Armée de l'Air Mirage F1Cs, flying in company with one of the later production Mirage IIIEs. (Dassault)

This photograph of a Mirage F1C-200 from EC 1/30 "Valois" shows that it is equipped with two Matra Super 530F-1 radar guided medium range air-to-air missiles, as well as a Barax jamming pod. This squadron, which was based at Reims is actually based at Colmar, with the designation EC 1/30 "Alsace" and equipped with the attack Mirage F1CT. (Sirpa Air)

EC 3/30 "Lorraine" is the French Mirage F1 OCU and has a complement of Mirage F1Bs and F1Cs, although recently its designation was changed to EC 3/33, being homebased at Reims. (Sirpa Air)

Two Armée de l'Air squadrons fly the recce/attack Mirage F1CR, both based at Reims, ER 1/33 "Belfort" and ER 2/33 "Savoie" (illustrated) both heavily involved in operations over Bosnia, especially the latest Operation Deliberate Force in August/September 1995. (Lert)

SPECIFICATIONS - MIRAGE 2000C

Wing: span 29 ft. 11.5 in.; aspect ratio 2.03; area 441.33 sq. ft.
Fuselage and tail: length 47 ft. 1.25 in.; height 17 ft. 0.75 in.; wheel track 11 ft. 1.75 in.; wheel base 16 ft. 4.75 in
Weights: empty 16,534 lb.; normal take-off 23,555 lb.; maximum take-off 37,478 lb.
Fuel and load: internal fuel 6,966 lib.; external fuel up to 8,200 lb. in one 343 U.S. gal. and two 449 U.S. gal. drop tanks; maximum ordnance 13,889 lb.
Powerplant: one SNECMA M53-P2 rated at 14,462 lb. dry and 21,384 lb. with afterburning.
Speed: maximum level speed clean at 36,000 ft. more than 1,270 knots.
Range: ferry range 1,800 n.m. with drop tanks; combat range more than 800 n.m. with four 551 lb. bombs or more than 1,000 n.m. with two drop tanks.
Performance: Maximum rate of climb at sea level 55,550 ft. per minute; service ceiling 60,000 ft.; climb to 49,000 ft., 4 minutes; time to intercept a Mach 3 target at 80,000 ft. from brakes-off, less than 5 minutes; take-off run about 1,500 ft. at normal take-off weight; g limits +9 normal and +13 ultimate.
Weapons: two internal DEFA 554 30mm cannon with 125 rounds per gun. Five underfuselage and four underwing pylons. Maximum weapon load 13,890 lb. Standard air defence load-out is two MATRA Magic II IR AAMs and two Super 530D radar guided AAMs. In ground attack role, up to 18 551 lb. bombs, as well as anti-runway bombs or laser guided bombs, can be carried, or AS30L laser guided missiles, or ARMAT anti-radiation missiles, or two AM39 Exocet anti-ship missiles.

Three generations apart, an Armée de l'Air Mirage IIIC from the now defunct EC10 flies leader to the second prototype of the Mirage 2000, also known as the "electric Mirage", due to its digital fly by wire system. (Dassault)

The Orange air base, in the south of France is the home for three Mirage 2000C/B (all with the improved RDI radar) squadrons, these are EC 1/5 "Vendée"; EC 2/5 "Ile-de-France" and EC 3/5 "Combat-Venaissin. (Sirpa Air)

A Mirage 2000C leads a pair of two-seat attack Mirage 2000Ns in desert camouflage. (Sirpa Air)

Mirage 2000C from Escadron de Chasse 1/12 "Cambrésis". (Author)

Close-up of a French Mirage 2000C with unusual nose, or intake art! (Author)

The latest Mirage 2000-5 has improved capabilities, including a new RDY radar, glass cockpit and the new Matra Mica AAMs, as illustrated in this two-seat development machine. (Soupart)

The Mirage 4000 was an unsuccessful attempt at making an scaled up Mirage 2000 with two engines, especially for the export market, Saudi Arabia almost bough a considerable quantity, but at the end acquired Panavia Tornados, both in the air defense and strike variants. Only this prototype was built. (Dassault)

(Above and Opposite) The Dassault Rafale, named after a wind, is the French manufacturer last fighter, ordered in quantity by the Armée de l'Air and the Aéronavale, will enter service in 1998, and although with a different name, for many it is a fourth-generation Mirage, it could not be the other way!. (Dassault)

Rafale
AMD-BA

2

Mirages in Israel, The Early Years

During the period 1962-1967, the early years of the Mirage III in Israeli service, it is difficult to disassociate this fighter from the rest of the Chel Ha'avir (Israel Air Force). Therefore, both here and in the Six Day War Chapter, the narrative covers not only the Mirage story, but also the Israel Air Force story during those turbulent years. This approach also allows the reader to better understand the special environment which surrounded them.

It was Saturday, April 7, 1962, when a pair of strange looking aircraft descended from their cruise altitude to land at Hatzor air base. Never before had the air and ground crews in the many air forces of the Middle East seen the delta shape of the Mach 2 fighter that was to become such a familiar sight in the score of years to follow.

Nevertheless, aside from the air and ground crews, the landing at Hatzor air base of the first Mach 2 fighters in the Middle East went almost unnoticed. No public announcement was made and no formal ministerial delegation, as was usual by then each time the first of a new type of combat aircraft arrived at Israel, came to greet the IDFAF (Israel Defence Force/Air Force) in its moment of glory. It was Saturday, the formal day off in Israel, and as such no official ceremonies could be held.

But to Rav Seren Dani Shapira, who led the first pair and was to lead — in their ferry flights — eighteen more formations of Mirages in the following four years, it was a major highlight in his long career as the IDFAF's Chief Test Pilot. During that Saturday in the early spring of 1962, Shapira was bringing to the Middle East the first of a new breed. The Mach 2 fighters were to rule the skies of the Middle East for almost fifteen years.

Although Shapira could not tell at the time, he was ferrying the most successful of this new generation of combat aircraft. The Mirage represented a quantum leap in capabilities over the previous generation of slightly supersonic fighters. In the coming years, the Israeli Mirages were to achieve a worldwide fame. The Mirages were evolved into a true multirole combat aircraft: they attacked and they bombed, they photographed and they attacked ships, they interdicted and they intercepted, but most of all they ruled the skies of the Middle East with several hundred confirmed kills to their credit.

It was not always like that. The beginnings were hard and painful. The first four years of service brought many disappointments from the revolutionary French fighter that saw its origins as a point defence interceptor.

On May 11, 1957, it was stated during the weekly meeting of the IDF's General Staff (which was attended by the Minister of Defence David Ben-Gurion) that France was offering Israel a new combat aircraft: the Dassault Mirage III. The Prototype flew on November 18, 1956 and was in fact the only existing example of the Mirage when the General Staff discussed the French offer, thus revealing the close ties between France and Israel at the time. However, almost exactly two years would elapse, following the original French presentation of the project, before an Israeli test pilot flew the type.

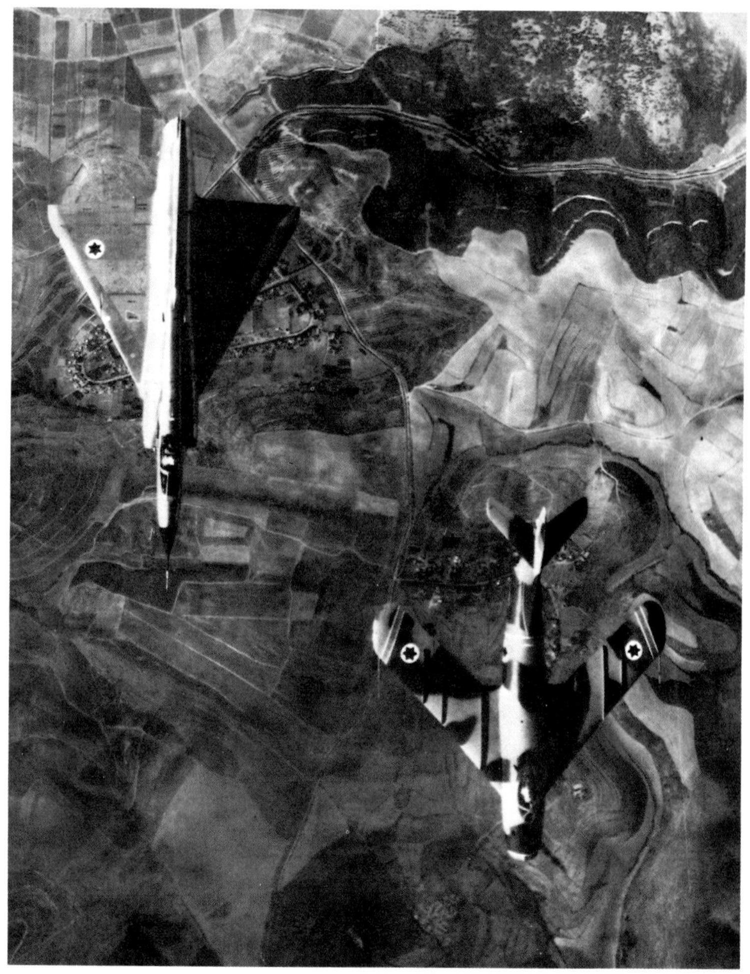

Arab air forces MIG-17 Frescos and Israeli Mirage IIIs have tangled over the Mid-East skies many times since the early sixties, until the October 1973 War. IDFAF (Chel Ha'avir) Mirage pilots admired its low speed maneuvering capabilities. This photograph shows a Mirage IIICJ flying escort to a former Syrian Arab Air Force MiG-17F "Fresco" whose pilot defected to Israel and sought political asylum. Note Israeli markings, while the recognition bands were applied by the Syrians. (IDF)

The IDAF delegation to the 1959 edition of the Paris Air Show was headed by Aloof Ezer Weizmann, but also included Sgan Aloof Moti Hod and Rav Seren Dani Shapira, who just had graduated from the French Test Pilot School (EPNER) at Istres. On June 23, Shapira strapped himself into the tight cockpit of the new interceptor (in fact it was the third pre-series development aircraft) while Wizmann and Hod carefully watched as the Mirage IIIA 03 departed. There were still no two seat conversion trainers as such and the type was still under test; Shapira was only given rudimentary instructions from Dassault's test pilot Roland Glavany. It was later claimed by some that the brief instruction was part of the selling effort, showing the Israelis that — in spite of its performance — the Mirage was a relatively simple machine. During that first flight, Shapira reached Mach 1.2. During his third flight in that evaluation series, he became the twelfth pilot in France to reach Mach 2.

While Shapira was flying the aircraft, Weizmann tested his own personal rule which stated that "a good looking aircraft must be a good aircraft." The Mirage was a beauty by all means. Thus, with Shapira's report and Weizmann's enthusiasm, the result of that evaluation was a decision to procure the new fighter as the best possible means to ensure that the IDAF would stay on the technological forefront. It was obvious that the Soviet MiG-21 was to find its way to the Arab air forces sooner than later. But two questions were left open at that stage: the quantity to be procured and the role to be fulfilled.

Ezer Weizmann formulated a requirement for 100 multirole Mirages at a package cost of 200 million dollars. "If I were a combat pilot instructed to attack a target in Cairo," argued Weizmann, "I would much prefer to penetrate with a Mirage even if it means one ton of bombs less to be dropped on the target." Weizmann was fascinated by the opportunity to procure an aircraft that could do it all. Forget the traditional separation of bombers and interceptors. Here was the perfect machine. It could penetrate at a very high speed, attack accurately the designated target, and make its egress at even higher speeds while at the same time defending itself against hostile interceptors. This was accomplished without the escort of the friendly fighters which make the planning and execution of such an attack mission so demanding and such a force consuming effort.

What Weizmann really saw in his mind was a formation of four Mirages, led by Rav Seren Ran Peker. The formation was to successfully attack the distant EAF (Egyptian Air Force) air base at Ardaka and shooting down three of the four defending MiG-19s that tried to intercept the attacking Israeli formation; or, Weizmann envisioned another formation of four Mirages, led by Seren Eitan Carmi, successfully attacking the distant SAF (Syrian Air Force) air base at T-4 while shooting down the two MiG-21s on CAP that tried to interfere with their offensive action. This vision was to turn into reality exactly eight years later, but at that stage Peker still was a young Segen, Carmi was still to win his wings and Weizmann was still to fight for his vision because the vision of one is not necessarily the vision of the other. Only in rare occasions does it becomes clear which vision is indeed the right one to follow. That case, Weizmann's vision, was one of those few and rare occasions.

Weizmann's demand for 100 multirole Mirages was too much for the other members of the General Staff to swallow, or for the senior officers within the IDAF, for that matter. They all claimed that the Mirage III was a dedicated interceptor, which it was at that stage of development. The French had designed it without even cannons but with a boost rocket and a single semi-active radar homing (SARH) Matra R530 AAM, that it was still not in service. Staff members and officers also pointed out that this Mirage was awfully expensive.

Thus, they argued, only a small quantity of 24 Mirage III interceptors, sufficient to equip one squadron and to minimize the development risks, should be procured. Additionally, they stated that a large quantity of less sophisticated attack aircraft, even the Douglas B-26 was mentioned in that context, should be procured — introducing the now familiar mix of expensive and cheap fighters. The sum to be saved was to be spent by the ground forces to procure much needed equipment such as modern tanks or AA guns. It was a fine and balanced proposition which eliminated the risk of ordering what was, at the time, still a non-operational aircraft which would have to be dramatically redesigned to meet the specification which Weizmann formulated.

Weizmann tried every trick in the book to persuade the decision makers that his vision was the right path to follow; however, he failed to break what he referred to as the "24 barrier," even when he scaled down his requirement to ninety Mirages. There were other areas in which Weizmann was struggling at the time. The most important of all, in retrospect, was the change of the IDFAF's public image from just another arm of the IDF into that an elite force. In the end, there were contributing circumstances that broke the "24 barrier," but above all else stood this tall energetic officer, who had a vision and struggled to turn it into a reality in his own special way. This struggle was as real as that of Israel's president, who strove for peace in the Middle East.

There were two contributing circumstances.

In January 1961 Rav Aloof Zvi Zur succeeded Rav Allof Cahim Laskov, the former commander of the IDAF, as the sixth IDF's Chief of Staff. Zur, a member of the Hagana underground movement since 1939 when he was sixteen and a regiment commander with the Givatti Brigade during the War of Independence, was convinced by Weizmann's passion and arguments. Quality of the air and ground crews had been Weizmann's baby for the past three years. Combine these quality crews with the best — or almost the best — multirole fighter in the world and a devastating force would emerge, the kind that was to win its world fame six years later. Furthermore, a huge arms deal with West Germany was formulated around that time as part of the German compensation for the murder of six million Jews in the Holocaust of World War II. This influx of arms and military supplies probably reduced the objections of the members of the IDF's General Staff to increasing substantially the already huge, in percentage, IDFAF budget. The "24 barrier" was broken and, although Weizmann did not receive the ninety Mirages he asked for, he was quite pleased with the 72 ordered. This number was later increased slightly to 76.

Upon receiving the Israeli order, Dassault gave the IDFAF aircraft the designation Mirage IIICJ. The J stands for Jüif (Jew-

Until the first F-4E Phantoms arrived to Israel in September 1969, IDFAF Mirages from 101, 117 and 119 Squadrons (the latter would eventually transition to Phantoms) were painted in the traditional air superiority natural metal scheme. The picture shows four Mirage IIICJs from the first unit to reequip with the type, 101 Squadron, identified by the red/white striped rudder and winged skull insignia. Today this squadron flies F-16C/D Fighting Falcons. (IDF)

ish). Marcel Bloch, who had been Jewish until changing his belief and last name after World War II (which he partially spent in a German concentration camp), seemed to neglect the fact that his Mirages were bound for Israel; later on, he continued this line of thought when the IDAF Mirage 5s were designated Mirage 5Js. But, whatever the origins of the designation were, there was no doubt that the Israeli Mirages were indeed different from the dedicated point defence interceptor that Shapira evaluated in the summer of 1959. They were much closer in concept to Weizmann's vision of a true multirole fighter bomber: an aircraft that could penetrate hostile territory, accurately attack its designated target, engage the enemy interceptors, and return safely to its home base. That was the concept and, miraculously, in the era of modern military aviation that concept turned into reality.

Dani Shapira was the pilot behind that metamorphosis. He was the one who conducted the tests with the Israeli multirole version, that huge deviation from the original French concept. The SEPR.841 auxiliary rocket engine was deleted and extra bomb racks were added while the Armée de l'Air adopted the IDFAF requirement for a pair of cannons to augment the single R530. The DEFA 552 30mm cannon with 125 rounds per gun was the preferred weapon.

While the Israeli internal debate was over and Shapira was testing the innovations, a last minute obstacle in the shape of the French Foreign Minister had to be overcome. The Foreign Minister objected to the deal. However, the personal approval of President De Gaulle ended that conflict. Once all obstacles — both in Israel and France — were cleared, the stage was set for the professionals to step in.

Even then, there was still one further debate to be settled. After all, each military decision is preceded by a debate since the stakes are usually high and the chances of success are slim. The debate revolved around which type of pilots to send for the conversion of the new aircraft type; an aircraft which was, by all means, something very different from anything seen in the Middle East up to that time. One school of thought promoted veteran officers with their vast experience as combat pilots and leaders. The other claimed that such veterans, who for years had flown the Spits and Mustangs, might find it difficult to adjust to the new Mach 2 fighter. This second school felt that young and promising officers, who only spent a short period with piston fighters, would adjust themselves more quickly and thoroughly. Also, the second school argued, the younger pilots would have a longer service period ahead of them when compared with the veterans, who were already looking for promotions to staff officers or base commanders. The final decision was a compromise, but the two officers that were selected to command the first two Mirage squadrons were indeed veterans.

Rav Seren Joe Allon was a graduate of the first flying course in August 1950 and he converted to jets three years later as the vicecommander of 117 Squadron. By October 1956 he was flying Mystères with 101 Squadron as a Seren. Before the end of that year, at the age of 27, Allon succeeded Moti Hod as the commander of 113 Squadron. After completing his first tour of command, Allon commenced a second tour with 101 Squadron, commanding the unit until it was disbanded as a Mystère squadron in anticipation for its reactivation as a Mirage squadron during 1961. Allon was selected to command the first Mirage unit: 101 Squadron.

The other veteran officer selected to command the second Mirage unit was Rav Seren Shamuel Shefer, who, just like Allon, flew piston fighters for several years before converting to jets. This was, in fact, the second time that Shefer was nominated as the future commander of 117 Squadron.

In the autumn of 1956 it was planned that 117 Squadron would convert from the Meteors to the Canadair Sabres. Shefer headed the delegation that commenced the conversion in Canada in September 1956. However, the Sinai Campaign forced the delegation back home, the Canadians canceled the deal, 117 Squadron remained with its Meteors, and Shefer flew Ouragans with 113 Squadron as a Seren during the Sinai Campaign. A year later he succeeded Menahem Bar as the commander of 109 Squadron, a post that he still occupied when it was decided to send him to

France to convert to the Mirage and to nominate him as the commander of 117 Squadron.

Meanwhile, the single seat Meteors were retired and the two-seat Meteors were re-allocated to the neighboring 110 Squadron until they were phased out during 1970. Gideon Magen took his wife along with him during the last flight of that type in IDFAF service that year. It must be emphasized that by 1961, the flying course syllabus changed to all jet with the French Fouga Magister; thus, the advanced stage on Meteors with 117 Squadron at Ramat David air base became unnecessary as the new fighter pilots were streamed directly from the Magister to the OTU on Ouragans with 113 Squadron at Hatzor air base.

The other part of the compromise involved the four officers that accompanied Allon and Shefer to France. All four of them were in their mid-twenties, all four of them had graduated from 12 Flying School during 1954 or 1955, and all four were promising young fighter pilots. Seren Amos Lapidot and Seren Ezra Aharon accompanied Allon, while Seren David Ivry and Seren Dror Avnery accompanied Shefer.

It is a remarkable fact that all six pilots fulfilled their promise, each in his own personal style. Allon went on to become a base commander at Hatzerim air base. Having achieved the rank of Aloof Mishne, Allon served as the Israeli Air Attaché in the United States. While performing his duties in the United States, Allon was murdered on July 1, 1973, probably by Palestinian terrorists.

Shefer also became a base commander, in Tel Nof, and retired from IDFAF service as Aloof Mishne. Avneri and Aharon continued to fly fighters with the IDFAF as reservists until retiring from active duty about a score of years later at the rank of Rav Seren (reserve).

It was Ivry and Lapidot that continued in service until reaching the highest post within the IDFAF (Lt. General), Ivry from 1977 until 1982 and Lapidot succeeding him from 1982 until 1987.

The Mirage conversion officially commenced on August 3, 1961 at Mont de Marsan. The process included a familiarization with the Mirage's exceptional performance, by the standards of these years. The extreme boundaries of the flight envelope were explored, including Mach 2 flight, a zoom to high altitude, and even a flight in which the SEPR.841 boost rocket was activated — although that feature was deleted from the Israeli aircraft.

Mirage IIICJ from 117 Squadron takes-off from its base, note censored badge and serial number a common practice during the sixties, most of this censorship was applied to the negatives. This Mirage is fitted with two 1300-lit drop tanks. (IDF)

Upon completion of the conversion on September 18, 1961, the six pilots returned to Israel to organize the units, translate the technical manuals from French to Hebrew, and to lay the foundations for the conversion courses at Israel in which they were supposed to instruct the bulk of Israeli Mirage pilots. Simultaneously, a delegation of ground crew was converted in France with roughly the same goals. Finally the day came, on April 7, 1962 Dani Shapira led a French ferry pilot to deliver the first pair to 101 Squadron at Hatzor air base. As already mentioned no government officials were present. To rectify that situation, a formal ceremony was celebrated during Sunday, April 8, when Dani Shapira made a demonstration flight, the first of many take offs by Mirages and their derivatives from the concrete runways of this ex-RAF air base.

The working-up period of the Mirage in IDFAF service was not an easy one. The Mirage proved to be a demanding aircraft, both in the air and on the ground. It was certainly not trouble free, at least initially. The most bothering malfunctions involved the SNECMA Atar 9B engine and the fire control system. The first mentioned malfunction caused several accidents due to a sudden in-flight extinguishing of the engine. This problem was solved after a surprising incident in which a Mirage made a perfect belly landing in a field after its pilot ejected, as will be described later. The second problem, with the fire control system, resulted in many frustrating moments for fighter pilots who caught Arab fighters in their gunsights but were unable to shoot them down. The true range to the target, obtained by the Cyrano Ibis radar, was corrupted by a transformation box between the radar and the Thomson-CSF sight. This problem was traced by a group of technicians led by Rav Seren Ephraim Ashkenazi, a fighter pilot and an electronics engineer. Fixing this problem resulted in a dramatic change in the outcome of the aerial combats in the Middle East.

Other malfunctions, less traumatic but still bothering, involved the clamshell nozzle of the Atar 9B — which sometimes stuck in the open position (a problem that was rectified years later after switching to the Atar 9C), problems with the caution lights of the hydraulic systems providing false warnings, problems with the nose wheel assembly collapsing without any warning on the ground, and problems with the Cyrano Ibis radar — which had a low MTBF and was prone to overheating.

But it was common for a new combat aircraft to introduce new problems. These problems are eliminated, usually only to be replaced by others, during the early service career of the aircraft. Indeed, as these malfunctions were treated, the Mirages became what Ezer Wizmann claimed they would be: true multirole fighters, the dream of every fighter pilot, the makers of aces.

However, future aces have to be trained first. This was the first task of the new Mirage units, while engaged in the working-up and prior to being announced as operational units. Although Weizmann referred to the Mirage as a multirole fighter, it was still regarded as an interceptor. The Mirage was also the most advanced weapon in the Israeli arsenal, and as such only the best fighter pilots were streamed to the Mirage squadrons, usually after a period with 105 Squadron equipped with Dassault Super Mystère B2 (SMBD).

The absence of a two-seat conversion trainer until late in 1966 was not unusual in the IDAF; after all, the Ouragan, Mystère, and SMBD had no such luxury either. However, that lack of a second seat resulted in a feeling of respect, or even fear, for the delta arrow. The very sensitive control system was the instigator of many hairy incidents, including a pilot who oscillated in his Mirage from take-off to 30,000 feet and a pilot who could not understand why his Mirage was rolling from side-to-side until he realized that his right hand on the stick was shaking slightly.

The behavior of the French delta fighter was especially dangerous at low level and slow speed when marginal changes in the angle of attack (AoA) resulted in a rapid slow down. Due to the enormous drag of the delta configuration, when either an increase in the AoA or a decrease in AoA during rapid acceleration was created, flying this machine became a sensitive and demanding operation for the pilot.

Yet another interesting item was the ADHEMAR system which was an AoA indicator. The indicator did not present the exact AoA but, instead, presented a color: green for normal operation, yellow for caution, and red for dangerous. The pilots were instructed not to reach the red, such high AoA usually resulted in a disturbed air flow on the front engine face, bringing the distortion coefficient to an unacceptable level, and usually resulting in a compressor stall which is not the preferred state in which to find oneself during combat. Even worse, at high AoAs the Mirage was prone to abruptly enter into a spin from which the only mean to recover was to pull the ejection seat handle.

However it seems that almost all Israeli Mirage pilots elected to ignore the ADHEMAR and flew the fighter well into the red, feeling the Mirage through the seat of their pants and accepting the risk of an engine flame out in combat. A nice story to illustrate that point is told about a young Mirage pilot who, after almost two years of service in a Mirage unit, was under the threat of being expelled from the squadron. The reason for this threat was that the pilot, although promising, was losing in almost every air combat training sortie. There was no room for such a pilot, an easy prey, in an interception unit. It was its pure luck that, when the first two-seat Mirage IIIBJ arrived in late 1966, the squadron vicecommander decided to take this young pilot along with him in the back seat for an air combat hop so he might learn a thing or two. To this young pilot's amazement, the whole combat was performed with the ADEHMAR showing pure red and his commander up front, who was responsible for the flight safety in the squadron, ignoring it altogether. Nothing was later said, it was something that everybody did but no one dared to discuss. However, having taken that lesson to heart, that particular young pilot continued to fly Mirages and became an ace with five confirmed kills.

Not that compressor stalls and engine flame outs were rare, but the pilots learned to live with them. In many combat stories, from the three wars in which the Israeli Mirages starred, the frustrations of re-starting the atar at a critical stage of the battle over enemy territory or the instant urge to aid and protect a fellow pilot in a nearby Mirage whose engine flamed out and turned him into a sitting duck target for hostile fighters, are common themes. In one air combat, for instance, Seren Yehuda Coren from 117 Squadron, who ended his fighter pilot career as an ace with eleven kills, had to re-start his engine twice over Iraq. Coren still returned home with two confirmed kills to his credit, his second and third.

In the end, it was the difference between the regular, good squadron pilot and the ace that allowed for successful outcomes in these dangerous engagements. Indeed, many Israeli aces owe their fame and lives to the exceptional behavior of the delta French fighter — even when the safety rules were ignored and the ADHEMAR showed pure red for more than a few seconds.

Once the first two squadrons were declared combat ready, it was only a matter of time before the pilots would persuade their commanders to look for troubles, or, in other words, to set ambushes for hostile fighters as did 105 Squadron under the leadership of Rav Seren Yak Nevo shortly after the SMBD was declared operational. But while the SMBD achieved its maiden kill fairly quickly, the Mirage's first three years as the top performance Israeli fighter were really disappointing. This period from 1963 to 1965 did not see a single kill credited to the Mirage drivers. During this same period, at least one Mirage was lost and another one severely damaged while engaged in operational activity. No doubt about it, the pilots and the aircraft were still drawing the learning curve of the Mirage's operational graph. The later results, as the many peaks in that graph would show, were still far ahead in time.

The operational career of the Mirage in IDFAF service began during the summer of 1963, slightly more than a year since the first two Mirages were delivered to 101 Squadron. The first true operational sortie was the interception of an USAF RB-57 on July 19, followed closely by the first air combat on August 20. This air combat was certainly not a preplanned ambush since Israeli ambushes at that period usually involved two pairs, the first acting as the bite and the second coming in for the kill. This time only a single pair of 117 Squadron Mirages from Ramat David air base, led by Seren Ezra Dotan, was scrambled to intercept eight Syrian MiG-17s over Lake Galilee. The battle was inconclusive, neither side losing an aircraft, and it might be assumed that both sides were not disappointed as the Syrian numerical superiority countered the Israeli quality edge.

Three months had to elapse before another interesting operational sortie was executed. It was November 11, when the two vicecommanders of 101 Squadron (Seren Ran Peker and Seren Amos Amir) were to test Weizmann's philosophy. By this time intelligence pieces consistently pointed out that the Egyptians were receiving and deploying Soviet SA-2 batteries, the first such weapons in the Middle East. It was decided to try to verify those claims with a deep penetration reconnaissance mission but the problem was that the then current IDFAF recce platforms, the Vatour IIBRs and a Mystère equipped with cameras, either lacked the required range performance or were unable to protect themselves against the Egyptian defenses (which by then included MiG-19 and MiG-21 interceptors and, possibly, the SA-2 batteries). The last recce sortie deep into Egypt had been flown almost two years before, during early 1962, by a Vatour IIBR at night. It was obvious that for a daylight penetration, a radical solution was required.

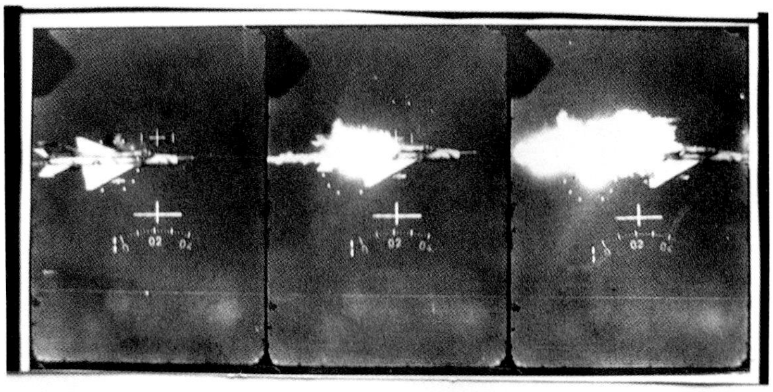

Difficult deflection shot! With an angle of 90 degrees this Israeli Mirage pilot shows that he learned his job very well while on the air-to-air gunnery training missions. This Egyptian MiG-21 was destroyed during the October 1973 War, as can be seen range was 200 meters. (IDF)

The solution found was neither radical nor new; it was simply the adoption of Weizmann's idea with one small difference: instead of bombing Cairo the requirement was to photo Cairo. Thus a standard Mirage IIICJ, number 53, was locally modified with a vertical low level (1,500 feet and 540 knots) camera of German origin. The camera, manufactured by Zeiss, was placed in the nose and the appropriate switches were mounted in the cockpit. The camera replaced the Cyrano Ibis radar in the nose but the standard armament of two 30mm DEFA 552 cannon was retained (see drawing).

The plan was that Peker, flying the modified Mirage number 53, and Amir, in a standard Mirage as an escort, would penetrate the Sinai at low altitude (100 ft. AGL), photograph El-Arish air base and Port Said harbor, and then turn into the Nile Delta to photograph the air bases and other military installations there, while also trying to photo the SA-2 batteries. Egress was planned via the Mediterranean Sea and then east to Hatzor air base. The plan was perfectly executed by the two pilots, both of whom were later to become Mirage squadron commanders and known aces, and even the EAF interceptors that were scrambled to intercept the pair had no chance due to the perfect planning and execution. However, the snag that almost ruined the sortie was not a mistake in the planning, not a lousy execution, and not even a lucky intercept by EAF fighters. It was a mere technical malfunction which, until then, had resulted in the lost of sixteen Mirages all over the world.

Just as Peker saw Hatzor air base, he watched the RPM and the engine temperature dials drop violently, immediately followed by an engine flame-out. Peker, with the important films just ahead of him in the nose of the doomed aircraft, zoomed for height; but, the Mirage, with its engine out, was sinking rate of 12,000 ft. per minute. It was surely not a glider and Peker knew that he could not make it to Hatzor air base. He tried, in vain, to re-start the Atar and then, instead of ejecting immediately, stayed in his aircraft to clear an agricultural settlement just below him. At an altitude of less that 500 ft., and only when he was sure that the Mirage would not crash into houses, he pulled the handle of the Martin Baker ejection seat and left the cockpit of that specific Mirage for what he probably thought was the last time. But, as Weizmann himself later put it in a song that he wrote to celebrate the occasion, the spirit of the pilot was shared by his aircraft and Mirage IIICJ number 53 made a perfect pilotless belly landing in a nearby field. As he reached the ground and witnessed the miracle, Peker immediately rushed to the Mirage and closed all the switches to prevent a possible fire.

Later on, the films were retrieved and the Atar 9B was run on the test bench. Only then was it discovered that the gaskets of the fuel injectors at the combustion chamber were unable to withstand the high temperature, well over 700 degrees centigrade, for an extended period. Thus, when the overheated gaskets failed, the fuel simply flew around the combustion chambers instead of being injected into them, and the Atar flamed-out. That problem was probably the most serious of the early malfunctions, causing the loss of French and Israeli aircraft, and it was discovered by pure luck. The epilogue of that story was that number 53 was rebuilt by the IDFAF Central Maintenance Unit (CMU) and Peker himself first flew the aircraft in 1965. Mirage number 53 was ultimately lost during the Attrition War of 1969/1970, as Israeli Consul Ran Peker was lost during the early nineties in Los Angeles.

Although not an operational activity, but also around that time, in August 1963, 119 Squadron at Tel Nof air base was disbanded. Its role as a night fighter unit was no longer valid as the Mirages were regarded as all weather fighter (although they were certainly unable to fulfill that title . . . but anyway they were better than the Meteors and Vatours). The Meteor NF.13s were retired, except for number 157, which continued to fly with Israel Aircraft Industries (IAI) until the early 1970s as a chase plane and to test the Gabriel SSM radio altimeter. The Vatour IINs were transferred as attack aircraft to 110 Squadron at Ramat David air base to act as lead aircraft during special attack operations. The leader was expected to benefit from the presence of an experienced navigator behind him. The other Vatours in the formation would be single seaters, a concept that proved itself quite remarkably in the coming years. The temporary disbandment of 119 Squadron, in anticipation for the delivery of more Mirages to complete the batch of 72 Mirage IIICJs, concluded in 1964, which allowed the formation of a third unit. The "Bat Squadron," as 119 Squadron is sometimes referred to, was re-formed as a Mirage unit early in 1964 under the command of Rav Seren Rafael Harlev, a veteran fighter pilot who flew with 113 Squadron as a Segen during the Sinai Campaign of 1956. Later, Harlev became one of the youngest squadron commanders in the IDFAF history when he assumed the command of a fighter squadron at the early age of 25.

While 1963, the first year of operations for the Mirage in Israeli service, passed without any major achievement except for Peker and Amir's recce mission, the second year, 1964, also proved disappointing in terms of air combat, although it also had its own first. During 1963 the Mirage made its first operational recce sortie. The first attack mission was achieved during 1964. On November 13, it was Friday for anyone who cares, a skirmish developed on the Syrian border between Israeli and Syrian ground forces and, for the first time since April 1951, the Israeli government decided to attack immediately Syrian artillery batteries and other military installations at the Golan Heights. The Golan Heights batteries and installations that shelled at the Hula and Jordan Valleys were roughly 2,400 ft. (800 m.) above the Israeli settlements.

The order was received at Ramat David air base with surprise but Rav Seren David Ivry, the newly appointed commander of 117 Squadron and the actual base commander in the absence of Sgan Aloof Yehezkel Somekh, immediately organized a task force. This force consisted of two formations of four Mirages from his own squadron, one to provide CAP cover and the other to conduct the maiden operational ground attack missions of the Mirage, plus two other ground attack formations of four aircraft each from the neighboring 109 Squadron with Mystères and 110 Squadron with Vatours. The only other significant contribution to the Ramat David effort was a formation of ground attack SMBD from 105 Squadron at Hatzor air base flying their first operational mission in their new role as ground attack aircraft. Their early role as interceptors had been take up by the Mirages. The action itself was successful. It was the first of many incidents in which the IDFAF was called upon to act as "flying artillery" to counter the Arab's numerical, and sometimes even technical, superiority in that field.

Although the aircraft attacked Syrian targets well into the Golan Heights, the SAF did not scramble its interceptors, so the CAP formation was left without any work. However, a strange incident involved the other Mirage formation, which was led by Seren Ezra Dotan. Dotan forgot that the Golan Heights is 2,400 ft (800m) above sea level and planned his attack as though he was attacking a target at sea level. As a consequence he barely managed to pull out of his dive before impacting the ground, punishing in the process himself, and the Mirage, with a very high G load.

It was the same month, November 1964, that the Israeli Mirages witnessed their second air combat. While in the first it was claimed that the numerical superiority saved the Syrian MiG-17s, no such excuses could be stated after a pair of Royal Jordanian Air Force (RJAF) Hunters photographed the two Mirages, led by Segen Oded Sagee, with which they fought. Only poor gunnery, or maybe a faulty fire control system just like that of the Israeli Mirages at that time, prevented the Jordanian pilots from translating their superiority into more noticeable results than the gun camera films that were published the next day in the Arab press. However, although the RJAF pilots could not know it at the time, they did cause the loss of an Israeli Mirage as the Israeli wingman ran out of fuel and had to eject from the powerless dart, the first of two such known incidents.

Yet one more air combat occurred on March 3, 1965, when two EAF MiG-17s met with IDAF Mirages without any results achieved by either side. Thus, by the year 1965, after the last of the 72 Mirage IIICJs were delivered, the three Israeli Mirage squadrons were already battle proven after performing operational recce and ground attack missions. Some of the bothering early malfunctions were already cured, most importantly the fuel injectors problem with the Atar 9B, and some, including the fire control problem, were yet to be analyzed and cured before conclusive results were to be claimed in the role for which the Mirage was designed: the air combat arena.

However before discussing the successful series of air combats which began in the summer of 1966, it is interesting, and important, to trace the development of the first Israeli Air-to-Air missile, the RAFAEL Shafrir (Dragonfly). The RAFAEL Shafrir began its life with the SMBD in the minds of its designers, but was mostly associated with the Mirage and its later Israeli derivatives. It is also necessary to overview the other developments within the IDFAF at that time scale, the period between 1961 and 1966.

The story of the RAFAEL (the acronym in hebrew of Rashot Le'Pitoch Amchai Lechima <Armament Development Authority>) Shafrir AAM began in late 1957. During the fifties the AAM was considered as the fighter's future weapon while the cannon was claimed to be outdated. This scheme might have been correct if the technology of fire control systems and guidance techniques had not been in its infancy at that time. Even at the present time, in the world of the microprocessor, this scheme is not completely trusted. All modern fighters retain the cannon as a logical fall back. However, four decades ago the promise of the new concept was such that many fighters were designed from the outset without cannons. Even the Mirage III received this "outdated" weapon only because of the insistence of the IDFAF and in fact all Israeli Mirage kills until July 1967, except one, were obtained by cannon fire. It must be remembered that the early examples of the Mirage's Soviet equivalent, the MiG-21, which was operated by the Arab air forces, had no internal cannon. Its air-to-air weapons consisted of only two AA-2 Atoll AAMs which were probably not too reliable, as were all AAMs at that time. As a direct consequence, this gave a distinct advantage to the Israeli Mirages and their internal cannons. Still, the IDFAF could not ignore the new weapon and by August 1, 1957, the procurement of the Nord 5103 radio controlled AAM was discussed in a meeting devoted to the subject in the Ministry of Defence. Yet it was clear that the Nord 5103, intended to equip the SMBDs, was inferior to the other contemporary short range AAMs, namely the U.S. AIM-9 Sidewinder (then GAR-8). Since it was clear that the U.S. was not willing to sell arms to Israel, RAFAEL offered by late 1957 to develop an IR guided short range AAM.

The RAFAEL offer was greeted at first with suspicion since they had no background in that field. Thus, by July 15, 1958 it was concluded in a meeting headed by Shimon Peres, then Chief Executive Officer of the MoD, that the assembly in Israel of the Nord 5103 would be pursued while a special committee was setup to evaluate the possibility to develop an indigenous AAM. More than half a year elapsed until the IDF Chief of Staff, Rav Aloof Chaim Laskov, concluded a discussion about AAMs procurement policy on February 28, 1959 as follows:

(1) RAFAEL will develop an IR guided AAM.

(2) A small batch of Nord 5103s will be procured pending the progress of the development at RAFAEL.

(3) a small number of SMBD pilots will be trained in France to operate the Nord 5103 but no investment will be made to facilitate training in Israel until the outcome of the development at RARAEL will be made clear.

The above summary was only partly executed. In April 1959 the full scale development of the RAFAEL Shafrir AAM commenced. At about that time 105 Squadron vicecommander, Seren Zoric Lev, and the Operations Officer, Seren Yalo Shavit, were sent for an AAM course with the Armée de l'Air at Algeria. The Nord 5103 was not procured and the first IDFAF interceptors to be equipped with AAMs were not the SMBDs but the Mirages and the AAM was neither the Nord 5103 nor the RAFAEL Shafrir, but the Matra R530 SARH AAM.

The development of the Shafrir was not abandoned however and the first trial launches from an aircraft (against flares) were conducted with a Vatour IIN during April 1962 (while engaged in those tests a Vatour IIN was lost due to a fire in a hangar on July 22, 1962). The Shafrir was not a copy of the AIM-9B Sidewinder, as many press reports later claimed, and although some of the principles used were similar, the components and the layout of the two weapons varied considerably. At one stage the French were interested in co-operating with RAFAEL in the development of the Shafrir but later they adopted the German produced AIM-9 as their standard short-range IR guided AAM. The Israelis, who could not make a similar decision, had to continue with the development of the Shafrir, which could only intercept targets that were not maneuvering and only if it was fired from behind the target at a range of less that three or four kilometers. Even then the warhead was too small to destroy a target unless a direct hit was achieved; an unlikely scenario as the seeker head was not sensitive enough to ensure a direct hit. In many cases the pilots saw the missile either guide itself directly into the hot ground or off in the direction of the sun, instead of chasing the enemy target. But, as already stated, the technology was at its infancy. For a first try, the outcome was quite reasonable. Moreover it was the only missile of its type available. Although the Shafrir was not a successful weapon, it gave the IDFAF fighter pilots the opportunity to adjust themselves to the missile age at the earliest possible period.

Thus, on November 4, 1962 the IDFAF ordered 200 production Shafrirs even before the results of the final series of tests (which commenced at France on February 18, 1963) were analyzed. During the test in France nine Shafrirs were launched, including two with telemetry equipment which fell to the Mediterranean Sea. Five were launched at an unmaneuvering CT-20 target drone (three passed near the target, one did not guide, and one guided to the target but disappeared) and two at a maneuvering CT-20, both of which missed. Sgan Aloof Benjamin Peled, who headed the Armament Development Branch at the IDFAF Staff at the time, concluded that the Shafrir proved its capability to intercept an unmaneuvering target when fired behind it from a range of up to three/four kilometers; however, the warhead was too small and a distant detonator with a long fusing time should be used to enhance the destruction possibility in case of a near miss. Peled recommended to fix those faults in the 200 production missiles on order but to take into immediate operational service on the Mirage the preproduction Shafrirs. That year, 1963, the Shafrir entered service even before the final development tests were completed and during 1963/1964 two more series of test firing, from Mirages, were conducted. The first test was conducted with firings at supersonic speed and the second against a ma-

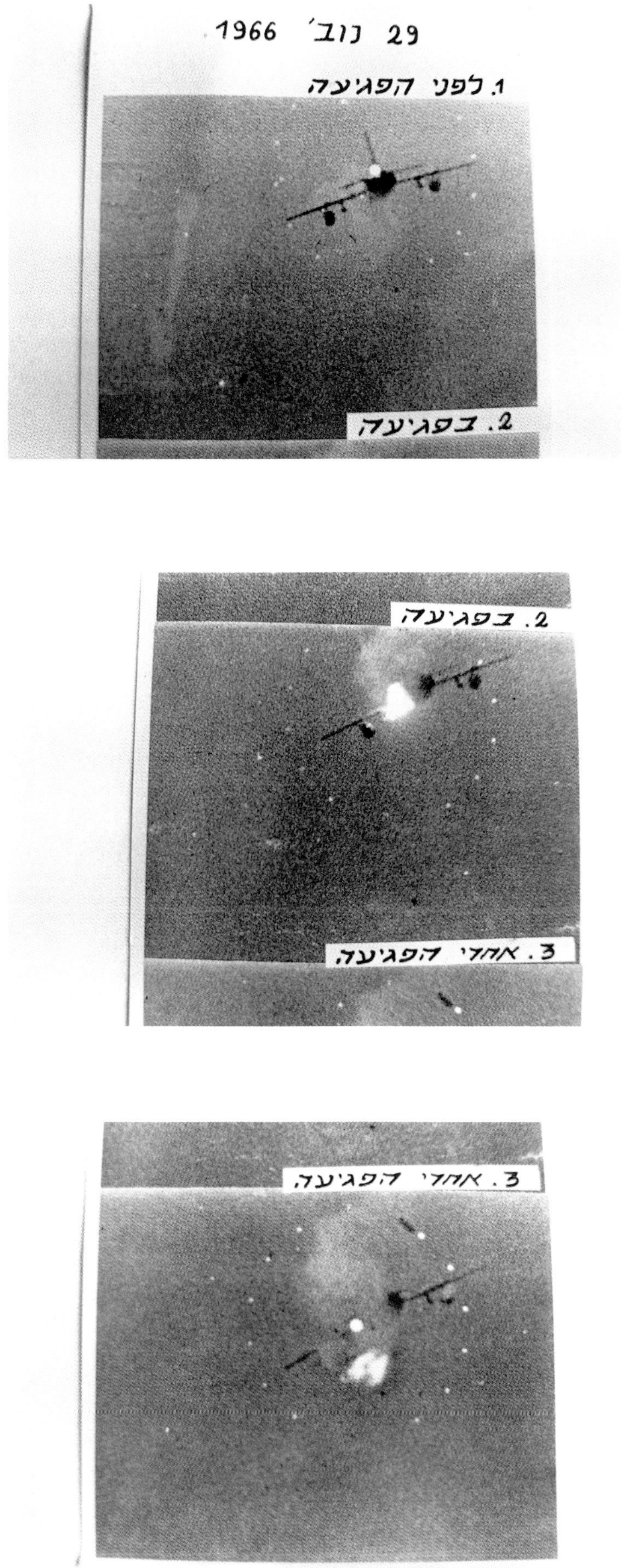

According to the Israeli Mirage pilot interviewed for this book, during the Six-Day-War air combats, both Jordanian and Iraqi Hunters were hard nuts to crack, this had a simple explanation, most of the pilots who flew these superb fighters had been trained by the RAF. (IDF)

neuvering target. Also in 1963 RAFAEL proposed to develop a new AAM, based on the lessons of the Shafrir program, and this was the more successful Shafrir 2.

Thus while the major milestones in the IDFAF history during the first half of the sixties were the introduction of the Mach 2 Mirage and the first steps toward the missile age, those years were also full of activity in many other aspects.

One of those aspects was the procurement activity. During that time scale, two major arms deals were concluded between Israel and nations which were not, until then, among its traditional arms suppliers. In addition to those two large arms deals, several other deals resulted in the introduction of new types of equipment on one hand and the augmentation of the already in service types on the other.

The first large deal was the supply of material by West Germany as part of the compensation for the Holocaust victims. The negotiations with West Germany had started in 1957 and by 1961 the first fruits of that effort arrived in Israel in the shape of Oerlikon L70 44mm radar-controlled guns. Although today the Air Defence Corps is an integral part of the IDFAF, at that time it was part of the artillery corps and the IDFAF had to wait one more year to receive its share in the German deal. The total gain to the IDFAF from late 1962 was some eighty aircraft, including sixteen Noratlas transports, 24 Sikorsky S-58 helicopters, Magister trainers and Dornier D-27 observation light planes (named "Dror" — Sparrow — in Israeli service). This was a massive addition of force to the then current units while, as an indirect consequence, resulting in the creation of several new units.

The sixteen Noratlas were absorbed by 103 Squadron and, with a planned force of 24 of the twin boom French transports, there was clearly no room in 103 Squadron for the faithful Dakotas. The Dakotas were separated from the unit to form 131 Squadron during 1963. By December of that year, 103 Squadron had on strength sixteen Noratlas at Tel Nof air base. The lack of space at this air base, which housed 12 Flying School, 113 Squadron, 116 Squadron, 119 Squadron, 124 Squadron, and the CMU, resulted in the relocation of 141 Squadron at Lod air base. Lod air base was opened on the western side of Lod International Airport (now Ben-Gurion International Airport) adjacent to the IAI facilities. At the same time, the ties with this company were tightened when the complete maintenance of the Dakotas was entrusted to its hands, relieving much needed work force within the IDFAF to more operationally important units and pioneering the current contract under which IAI maintains all the IDFAF transport fleet.

Yet another Tel Nof air base resident unit to enjoy the German deal was 124 Squadron. They received twenty-two H-34G-III and two SH-34Gs, which were supplied between December 25, 1962 (when the first four arrived by ship at Haifa port) and August 1963. With such an addition, the two S-55s were withdrawn from use and the helicopter pilots training scheme had to be revised. Until then most of the Israeli helicopter pilots were graduates of 12 Flying School which were converted to rotary flying in France or Germany. There was no helicopter pilot stream within the Flying School. That approach was reasonable when the IDFAF helicopter fleet was relatively small with about ten on strength, but was completely unrealistic when that force was more than tripled almost overnight. The obvious solution was to establish a helicopter pilots' stream within 12 Flying School. During 1965, thirteen Bell 47 and AB-47 helicopters were procured from French surplus stocks to fulfill the requirement for primary trainers. These were not, however, strictly allocated to 12 Flying School, but were taken up by a new squadron at Tel Nof, 123 Squadron. 123 Squadron shared its helicopters with the Flying School for instructional use but also operated the Bells, and the small number of Alouettes which had been relocated to the new unit from 124 Squadron, for operational activity in the missions of liaison, observation, command post and CASEVAC. The future helicopter pilots were thus trained, within 12 Flying School, at first on the Bells of 123 Squadron and then on the Sikorskys of 124 Squadron.

The final aerial elements of the German deal included the Drors and the Magisters. The Drors entered IDFAF service during 1964 and were allocated to 100 Squadron, to supplement the faithful Piper Cubs. The Drors proved themselves as extremely efficient in the observation task with the large side windows which allowed the unobstructed use of a 20 x 120 binoculars. In the same way, the Magisters were also used to supplement a current unit; they were allocated to 12 Flying School. During 1964 an emergency unit was formed within 12 Flying School, succeeding in many ways except the designation the days of 116 Squadron with Mustangs. (*This sentence is unclear, please clarify.) This unit operated armed Magisters, which were regular training aircraft and acted in this role for most of their service career. They were secondarily tasked with the CAS (Close-air-Support) mission.

The Magisters were armed with two 7.7 mm Browning machine guns in the nose and either 80 mm rockets or 50 kg light bombs under the wings. The Magisters commander post was covered by the C.O. of the Flying School's Advanced Squadron. Thus, while from the German deal the support elements of the IDFAF were the main beneficiaries, the American deal resulted in the creation of a new element within the IDFAF which was, in retrospect, a first move toward the amalgamation of the ADC and the IDFAF years later.

The requirement for reliable SAMs resulted from the realistic fear that the Tu-16 medium bomber might be supplied to the Arab air forces. Although modern interceptors like the Mirage III could tackle that threat by daylight, it was not that obvious whether or not a successful interception at night was feasible. Additionally, in certain operational scenarios the interceptor force might be needed elsewhere, or even temporarily exhausted by enemy action, so a last line of defence was seen as essential.

The only Western SAM system which promised to fulfil the Israeli requirement was the Raytheon MIM-23 Hawk which had only one minus: it was a U.S. product. Up until then the U.S. refused to supply arms to Israel although it encouraged its Western allies (especially France, but also Great Britain and Germany) to do so in order to preserve the military balance in the Middle East. However, the Hawk system was certainly not an offensive weapon and, thus, the Israeli Primer Minister and Minister of Defence, David Ben-Gurion, could rightly request from President Kennedy the supply of the Hawk as a strictly "defensive weapon system."

As usual with the supply of U.S. weapons, the procedure is quite prolonged; but, on November 27, 1962 the Kennedy Administration approved the Hawk sale to Israel. This was the first time ever in which the supply of a modern major weapon system to Israel was approved by any U.S. administration. The approval was followed by a debate within the IDF as to who would control and operate the new weapon. The ADC, while part of the Artillery Corps, was used to operate only AA guns and the Hawk was certainly more of an air vehicle than an artillery shell. Moreover, since the Hawk batteries were to cooperate closely with the IDFAF interceptors and with its early warning command and control network, it was only natural that the new weapon would be allocated to the IDFAF.

The debate ended with the decision to allocate the new weapon to the IDFAF. This was the first step toward the complete amalgamation of the ADC within the IDFAF, less than ten years later, during the Attrition War of 1969/1970.

By early 1963 an Israeli delegation visited the U.S. to discuss the new deal and shortly afterwards went the personnel to convert to the Hawk, consisting of men from both the ADC and the IDFAF — although those from the ADC were formally transferred to the IDFAF. Training commenced at Fort Bliss with the battery training phase, which concluded in October 1964. Five months later, in March 1965, the first five batteries arrived at Haifa. The weapon was shown in public for the first time during the 1965 Independence Day celebration.

As was usual then, the introduction of a new weapon was followed by an "operational evaluation." The Hawk was no exception, but the missile which was launched at a Syrian fighter missed its target and the maiden kill of the weapon was to be achieved only during the Attrition War. Nevertheless, the introduction of the Hawk, the second such weapon in the Middle East after the SA-2 was introduced in Egypt two years earlier, was important. It marked the first step toward entrusting all types of air defence in the hands of the IDFAF. It was also important because it was such a valuable addition to the IDFAF interceptors in the defence of highly important military and civilian installations and because it was the first sign of the later obvious shift from reliance on French equipment to the current total Israeli reliance on U.S. weapons.

Another step in that direction was the procurement of long range heavy lift transports to restore the capability which was lost when LATA was disbanded during the closing stages of the War of Independence. Again, U.S. equipment was used because no French equivalent was available. The preferred heavy lift transport of the IDFAF and its commander, Aloof Weizmann, was the Lockheed C-130B Hercules; however, the 1960 IDFAF proposal to acquire two C-130Bs was rejected since the Hercules was too expensive and its availability was not assured. Thus, the Israelis had to set their sights lower at the alternatives which IAI offered.

Both alternatives involved the procurement of surplus civilian airliners on the free market, their complete overhaul at IAI, and their conversion to a military standard. The two types that IAI offered were the Lockheed Super Constellation and the Boeing Stratocruiser. The first was quite a popular airliner during the fifties, before the first generation of jet airliners was introduced, and thus enjoyed the benefits associated with widespread use — although its military development potential was not overwhelming. The other contender, Boeing's Stratocruiser, was much less successful as a civilian airliner; however, it had similar variants, the C-97 Stratofreighter and the KC-97 Stratotanker, which meant that it had a military development potential.

Unable to obtain the preferred C-130B, the IDFAF settled for the Stratocruiser and, by 1962, IAI procured five examples which had been phased out of Pan American service. The five Stratos, as the type was to become known in Israel, were ferried to Israel during 1962 by IAI test pilot Hugo Marom. Work on the overhaul and conversion began immediately. In addition to the regular features in a military transport, such as the strengthening of the floor or the installation of paratrooping (originally said: paradropping) equipment, the IDFAF requirement included the installation of a moving tail to allow the ingress and egress of outsize cargo items. However, the engineering capability and experience at IAI at the time was quite restricted and it was revealed that the design was totally inadequate. As a result, only two of the Stratos were converted to include the moving tail. Those two Stratos suffered; severe limits were imposed on their flight envelope and, to the IAI engineers' embarrassment, once the tail was opened it was quite hard to close due to the deflection caused by the tail's own weight (which was miscalculated by the engineers).

While the Stratos were being overhauled and converted at IAI, an echelon of air and ground crews were converted to the type in the U.S. during 1963. On July 1, 1964, the Stratos were officially taken on charge by the IDFAF and were allocated to 131 Squadron at Lod air base. They were operated alongside the Dakotas and, as with the Dakotas, were maintained by IAI under a commercial contract. In the coming years the Stratos were to fulfill the original requirement for long range heavy lift capability stemming from the need to transport arms and equipment procured from abroad to Israel and, conversely, to transport military equipment from Israel to be overhauled abroad and returned. Indeed, just before the Six Day War and during the Yom Kippur War Israeli, long range transports established an air bridge with much needed items supplied by friendly nations.

Also, the IDFAF procurement effort at that time included the purchase of additional quantities of Ouragans and Super Mystères, which had been declared surplus to Armée de l'Air requirements. A dozen SMBD were procured in 1963 and allocated to 105 Squadron. The 105 grew to become the largest IDFAF fighter squadron by June 1967. While the procurement of the SMBD was well justified, the need for additional Ouragans was not quite obvious. Since the Sinai Campaign, 113 Squadron at Hatzor air base operated as an Operational Training Unit (OTU) and the Ouragan was regarded as an obsolete warplane whose days of service in the IDFAF were numbered. It was Rav Seren Zoric Lev that changed that view when he was posted as commander of 113 Squadron. Lev persuaded the IDFAF Staff that the Ouragan was an excellent ground attack platform and that the availability of surplus French aircraft at low prices made them highly attractive. Thus, by late 1964, 25 Ouragans were procured from France. No payment as such was charged by the French. The Ouragan were manufactured under U.S. financial aid and, thus, the French told the Israelis to pay the Americans, but the U.S. representatives simply refused to be paid by Israel for French arms and the Ouragans were in this form received free! By December 1964 all 25 Ouragans had been ferried to Israel and a new fighter unit was formed at Ramat David air base, "The Lion Head Squadron."

So by the mid-sixties the IDFAF fighter squadrons were equipped with French fighters only while the support elements were equipped with French, American, and German equipment. The early sixties were not only important because of the introduction of the Mirage, which was discussed in detail earlier in this chapter, but also because of the vast improvement in the transport and helicopter assault elements.

Those years also saw the introduction of the Hawk batteries and modern early warning radars of American origin. America was also the source from which Aloof Ezer Weizmann obtained his last material contribution to the IDFAF before leaving its command after eight long and extremely successful years.

However before Weizmann's visit to the U.S. is discussed, it should be remembered that the IDFAF did not abandon its French supply lines. During 1965 two large procurement contracts were concluded. One of those contracts was an order for five Super Frelon helicopters to add heavy lift capability to the helicopter fleet. The Super Frelons were delivered from April 1966 onwards and were allocated to a newly activated unit under the command of Rav Seren Chaim Nave. The other contract was for fifty Mirage 5s (a simplified interdiction version of the Mirage IIIE) which were to be supplied to Israel from late 1967. Yet another Israeli request, for Dassault Mirage IV bombers to replace the Vatours, was turned down by the French.

The change in U.S. policy became apparent during 1963. The American shift from indirect to direct involvement was due to U.S. considerations from which Israel benefitted. One of these considerations was the support of King Hussein of Jordan and it was the U.S. decision to supply Jordan with modern weapons, including Lockheed F-104 Starfighter interceptors and M-48 main battle tanks, which led to a compensation deal being discussed with Israel. It was that background which led Israeli Prime Minister Levy Eshkol to ask for the supply of U.S. weapons to Israel when he met President Johnson in his visit to the U.S. in June 1964. One of the weapons systems Prime Minister Eshkol mentioned was the Douglas A-4 Skyhawk.

In the discussions that followed, it became apparent that the bulk of the deal would involve the procurement of aircraft. Weizmann was instructed to write down Israel's shopping list. That shopping list included 210 aircraft of two types — the A-4 and the Grumman A-6 Intruder, both were attack aircraft which were operated by the U.S. Navy. The Israeli delegation to the U.S. was headed by Weizmann himself and Sgan Aloof Benjamin Peled, who was by then the commander of Hatzor air base. During that visit, in October 1965, the U.S. rejected the request for A-6s, as well as turning down the Israeli interest in the McDonnell F-4 Phantom; however, it resulted in a U.S. conditional agreement to supply Israel with 48 A-4s, the condition being that the French were indeed unable to provide with additional Vatours. By early 1966 it became apparent that the Vatours were not available and the U.S. formally announced the Israeli contract for 48 A-4Hs, as the Israeli subtype was designated, to be delivered from late 1967 onwards.

It was April 27, 1966 when Aloof Weizmann left the IDFAF after eight years in command and eighteen years of service, during which he was gradually promoted from a regular fighter pilot to the top of the air force. However, Weizmann did not leave the IDF, as he was promoted to become the Vice Chief of Staff, a post he occupied until December 1969 when he retired from the military. During his eight years of command, Weizmann's contribution was enormous and in many ways it was he who turned the IDFAF the deterrent force it is today. He emphasized the human factor and made the air force a professional elite outfit, a "winning force" as he like to call it. It was under his leadership that the IDFAF went beyond Mach 2 and obtained a substantial transport capability and, for the first time, a true vertical assault option. Simultaneously Weizmann did not neglect the training element, which entered the jet age also, and the early warning, command and control elements that were modernized while the Hawk batteries were added as another link in the air defence network. In addition the IDFAF received a major boost by the construction of Hatzerim air base in the Negev, which was the first IDFAF air base that was not of World War II RAF heritage but wholly designated by the service in line with its specific requirements. Hatzerim had its origins in a requirement that was formulated during Aloof Tolkowski's period of command, precisely on May 2, 1957. However, it was during Weizmann's period that the scheme was approved and construction commenced. The new air base was formally opened in 1966 with Sgan Aloof Jose Aloon as its first commander. The first resident unit relocated was 12 Flying School, under the command of Rav Seren David Ivry, transferring from Tel Nof. Shortly afterwards, 12 Flying School was followed by 123 Squadron, vacating much needed space at the crowded ex-RAF air base.

Eight years are not a short period by any standard and the five years which Aloof Mishne Moti Hod spent as the IDFAF Vicecommander were not a short period either. Hod was Weizmann's favorite successor but the Generals of the IDF General Staff though otherwise and thus, around 1964, Weizmann decided to continue in his post as the air force commander until convincing them. Finally a compromise was agreed. Hod was indeed nominated as the seventh IDFAF C.O. but he was not to be promoted to Aloof, which meant that he was under trial and, in retrospect, no one had foreseen what a successful trial it was to be. While of the first six commanders of the IDFAF three were not airmen and three were the graduates of RAF World War II vintage, neither can be said about the seventh commander although, unlike his successors, he was not a graduate of 12 Flying School.

Moti Hod was born in 1926 as Moderchai Fine at Kibbutz Dganya, the kibbutz just south of Lake Galilee which held against several Syrian armored attacks during the War of Independence in 1948. Eighteen years later, by late 1944, he volunteered to the British Army and served in a transport unit in Italy until the end of World War II. Then he stayed at Europe as a member of the Hagana underground organization. He played an active role in the smuggling of Jewish refugees to Palestine from 1945 until 1948. In 1948 he decided to become a pilot and commenced a private flying course in Italy. From there he was sent to Czechoslovakia and, by December 1948, with slightly more than 100 flying hours to his credit — of which only seven or so were on the Spitfire — he ferried a Spit back to Israel in Operation Velveta 2.

Back in Israel he was chosen, together with three more young Israelis, to attend an advanced training stage which was organized within 101 Squadron under the command of the legendary Rudy Augarten. The four graduated the course on March 14, 1949. Since no OTU was available yet, Hod joined 101 Squadron as a fighter pilot and by March 1950, exactly one year after receiving his wings, he was sent for conversion to jet fighters with the RAF to become the first Israeli jet pilot. After returning to Israel he assumed the command of 101 Squadron, on Mustangs, as a Seren and graduated from the IDFAF Staff College during 1953. That year he also married his wife, Pnina.

As a Rav Seren, Hod was posted to the Headquarters of Wing 1, Ramat David air base, under the command of Aloof Mishne Weizmann. During that period, he was converted to the Meteor with 117 Squadron. By March 1956, Rav Seren Hod succeeded Seren Benjamin Peled as 113 Squadron C.O., a post which he occupied until after the Sinai Campaign when he was promoted to Sgan Aloof and appointed the C.O. of Wing 4 at Hatzor air base. Leaving the first units for the first time, Hod was appointed a Head of Branch within the IDFAF Staff before he was nominated for the post of Head of Operations Department. In fact, he was appointed as the IDFAF Vice C.O., during 1961, where he stayed during five long years until becoming the seventh IDFAF C.O. on April 27, 1966.

During his service as a staff officer Hod insisted on flying the IDFAF latest fighters. He was converted at first to the SMBD and then to the Mirage, flying from time-to-time with the combat units. He won the admiration of the pilots with his exceptional skills and gunnery results.

As already mentioned, the trial period of Moti Hod, which was originally set to last one year, was to prove extremely successful and it was less than three months after his nomination as the IDFAF C.O. that Hod returned the first dividend to those who put their faith in him. On July, 14 1966 the IDFAF mounted its largest effort to that point to destroy the engineering equipment and infrastructure of the Syrian works.

The Syrians were attempting to divert the course of the Jordan river, eliminating Israel's fresh water supply. Beginning that morning, the Mystères of 109 Squadron and the Vatours of 110 Squadron attacked the Syrian targets with the Mirages flying continuous CAP to cover them. Since it was a preplanned battle day, it was possible to call back to the squadrons some of the pilots that were on other duties — such as instructors at 12 Flying School or staff officers — but who had retained their skills with the combat units on Emergency Posting (EP).

It was not until noon that the SAF tried to intercept the attacking aircraft and pure luck played into the hands of Seren Yoram Agmon, an instructor at 12 Flying School who was called that morning to his EP with 101 Squadron. Agmon was number four of the CAP formation that took off at 14:00 hours to relieve another Mirage CAP quartet that had ended its missions. What followed was a perfect GCI interception.

As the four Mirages patrolled the area, the controller ordered their leader to fly west, strictly opposite to the direction from which the Syrian interceptors were expected to appear from, and accelerate with full afterburner. It was a classic GCI action: inviting in the Syrian fighters as though the Israeli CAP was withdrawing while at the same time were improving the four Mirages' energy state. It was up to the controller then; it seems that he or she was working just fine. Shortly after the formation was ordered to turn east through 180 degrees, Agmon announced "Tally Ho" on a pair of Syrian fighters flying southeast at low level, being given the lead by the formation leader. At 2,000 meters Agmon recognized the pair as MiG-21s but shortly afterwards the Syrian pilots broke hard, in fact so hard that Agmon was surprised, with one of the MiGs disappearing. Agmon went up into the vertical to kill off his excessive speed and then closed on the MiG-21, firing his first burst without any apparent results, from 350 meters and his second from 250 meters. The MiG was hit at the right wing root and, following an explosion, the right wing was torn-off and the Syrian pilot ejected from his spinning aircraft.

The Mirage's fire control problem was obviously over as was the contest. The Mirage unit of the 101 Squadron scored the maiden kill. The importance of that first victory was enhanced by the fact that the victim was a MiG-21, the most modern opponent of the Mirage in the Middle East at the time. The event was also the first time in which a kill marking was painted on the left side of the nose of any Israeli Mirage. It seems that nobody dreamed, at that time, that the Syrian roundel painted on the nose of Mirage number 159, in which Agmon scored his kill, would be joined by twelve more to rise that particular Mirage into the position of the top scoring Israeli Mirage, a position it shared with number 158/458/758.

Although the contest as to which unit would score the first kill was over, by year's end all three Mirage squadrons were credited with kills.

The second unit to score was 117 Squadron which achieved its third victory, the first two were scored over EAF Vampires when the squadron flew Meteors during 1955. The 117's score took place on August 15, 1966. On that day a small Israeli patrol boat that had beached at Lake Galilee was attacked by Syrian artillery fire, followed by an air raid performed by four MiG-17s escorted by a pair of MiG-21s. The four attacking MiG-17s launched rockets on their first pass and then entered a second strafing pass during which one of the Syrian fighters was downed by small caliber fire, diving into the lake with the pilot still in his cockpit. Meanwhile a pair of 117 Squadron Mirages from Ramat David air base were scrambled to intercept but failed to nail the MiG-17s, meeting instead the Syrian CAP. In a very short action the Israeli leader, Seren Yehuda Coren, shot down one of the MiG-21s to win a second victory for the Mirage within slightly more than a month. Moreover the Syrian action did not go unnoticed.

Shortly afterwards the 110 Squadron Vatours attacked the Syrian positions, both in retaliation and to ease the pressure from the Israeli ships. Although the skirmish of 15 August was highly successful as far as the IDFAF was concerned, it was overshadowed by the events of the coming day, August 16, 1966.

On that day an Iraqi pilot, Captain Monir Rdfa, took off on a routine training mission in his MiG-21-F-13 Fishbed C interceptor, number 534. Strangely enough, this time the Iraqi Captain took along with him the technical manual of the MiG-21 and, even stranger, he left Iraqi air space and entered Jordan, flying west toward Israel. Jordanian fighters tried to intercept the Intruder but the Hunters were certainly unable to catch a MiG-21 flying fast at 30,000 ft. during the short time required to cross Jordan. Then as the highlight of the mystery, as the MiG-21 crossed the Israeli border, it met a pair of Israeli Mirages that not only did not try to intercept the Iraqi MiG but even greeted the alien and welcomed him to join them. Moreover, in the cockpit of these Mirages were not the regular squadron pilots expected to be found there in a routine scramble.

The leader of the pair was Rav Seren Ran Peker, C.O. of 119 Squadron. His wingman was no other than Sgan Aloof Shamuel Shefer, the Tel Nof air base commander. The identity of those pilots makes several important points quite plain: first, the Iraqi pilot was expected; second, no mistakes by eager young pilots were to be understood or forgiven while the experience of the veteran leaders was to help them in case of an unexpected surprise; and finally, third, the high rank of the officers involved might provide the clue that the operation was top secret.

However. it is impossible to preserve secrecy when a Mirage, flown by Peker, leads in a MiG-21 to land at Hatzor air base while another Mirage, flown by Shefer, keeps an eye from above. Such a scene, witnessed by many of the thousands of military personnel of which an air base was made up, just could not be kept as a secret. Indeed, it was the honor of Aloof Mishne Hod to announce in a press conference that Israel had put its hands on an example of the Soviet MiG-21 interceptor. Hod revealed that the Iraqi pilot defected to Israel by his own will, after his family left Iraq for a secure place, but added nothing more.

This was the second time in which an Arab pilot defected to Israel with his aircraft. The first time was when Mahmoud Abas Hilmi, an Egyptian pilot, flew from El Arish air base to Hatzor in his Yak-11 trainer on January 19, 1964. Since the Yak was not a modern first line interceptor, that incident was not given the exposure of Radfa's defection; but, unlike the Radfa case, that first defection ended in tragedy when Egyptian agents murdered Hilmi, who previously had received a new identity and left Israel for a land of his own choosing, Argentina.

Since the MiG-21 was the major, and most dangerous, opponent of the IDFAF, it was only obvious that an evaluation of the Soviet interceptor would be imminent. By September 1966, after he was ground instructed by the Iraqi pilot, Sgan Aloof Dani Shapira, still the IDFAF's Chief Test Pilot, took the MiG-21 for its first flight in its new guise with the number 007 and the Israeli insignia. At first the flight envelope of the MiG was explored and then mock air combats were flown against all the Israeli types of fighters. This dissimilar air combat training (DACT) series included sorties against all the squadron commanders, vice squadron commanders and senior leaders of all the IDFAF frontline units. It was an effort which was to bear fruit, especially during the next year as in the later stages of the Israeli-Arab conflict most of the Israeli leaders had already acquired invaluable combat experience.

The conclusions of that exhausting effort were that the MiG-21 was a very good fighter indeed, superior to all Israeli fighters except the Mirage. As for the Mirage, it was plain that the two types were very much alike and the outcome of the DACT was often a direct result of the pilot's skill rather than any technical superiority. The claim that if the Israeli pilots were flying the MiG-21 and the Arabs were piloting the Mirage, that the results of the air combats would not be any different was not made for propaganda. It was a true belief on the Israeli side based on the experience gained in the series of DACTs just mentioned. Moreover, although the MiG-21 had many deficiencies compared to the Mirage, including its lack of control sensitivity and the pilot's poor field of view, it was a simple, robust and reliable design.

Even before the Israeli evaluation of the MiG-21 began, several Western countries showed interest in the new acquisition; however, due to political considerations (Israel and the USSR still had diplomatic ties at the time), the Israeli government refused to share that surprising informational gift with nations that usually refused to supply Israel with the much needed arms required to guarantee its survival in the hostile environment of the Middle East. Only after the Six Day War was the MiG-21 transferred to the U.S.

Meanwhile, as the Fishbed was evaluated in Israel, the series of air combats that had begun on July 14 continued and its was the turn of 119 Squadron to score the third kill for the Mirage on November 13. During that day the IDF conducted Operation Magresa, a large scale retaliation against Jordanian targets in Samoa, south of Jerusalem and Hebron, after three Israeli soldiers were killed after a military vehicle was struck by a mine. The operation involved a relatively deep IDFAF involvement including the attachment of a Forward Air Controller (FAC) to the ground forces command vehicle. The operation started on 06:15 when Israeli tanks shelled the Jordanian positions. Exactly at that minute a pair of Ouragans armed for CAS or interdiction, plus a pair of Mirages as CAP, took off. The Ouragans patrolled the roads to the area and by 06:30 they reported that Jordanian reinforcements were arriving. Their proposition to execute an interdiction was turned down because it was the intention of the command to keep the action as local as possible. At about 07:30 the first pair of Ouragans was replaced by a second but, a quarter of an hour later, four RJAF Hunters arrived in the area.

The appearance of the Jordanian fighters resulted in an order to the Ouragans to withdraw and the scramble of four Mirages to augment the pair of CAP which were then suffering from a very low fuel state and had to return home as the fresh Mirages arrived. The Jordanian Hunters were intercepted over the Yehuda desert and a violent air combat developed in which the Israeli pilots met with aggressive and skillful pilots that squeezed the most from their mounts — which were excellent fighters at low altitude. After six long minutes, three of the Jordanian Hunters managed to disappear while the fourth, piloted by one Mawafak Salti that was trained by the RAF, tangled with a pair of Mirages from 119 Squadron, lead by the unit's C.O. Rav Seren Ran Peker and his wingman Segen Giora Oren.

Peker chased the Jordanian Hunter into a canyon at very low altitude, with Oren keeping a watchful eye from above. It was two

In order to mark the conversion of 101 Squadron to the Mirage, this unique taxiing was performed, to mark the transition from propeller fighters, the Spit, to Mach 2 interceptors, the Mirage IIICJ. (IDF)

and a half minutes before Peker managed to damage the Jordanian aircraft with a burst of cannon fire. The Jordanian pilot decided to eject. The cockpit canopy was jettisoned; but, the moment the rocket of the ejection seat was fired the uncontrollable Hunter rolled through ninety degrees and the unfortunate Jordanian pilot was fired at the wall of the canyon in which the chase ended. Years later, the RJAF air base at Azraq was named after Salti. Salti's British widow, whom he met when he was trained by RAF, sent a letter to the IDFAF in which she asked to meet the Israeli pilot that shot down her husband. That wish was turned down due to security considerations.

Although that air combat was the highlight of the air action that day, it was certainly not the end of it as by 07:55 the third Ouragan formation took off. While patrolling the Jerusalem-Hebron road at low altitude, due to a restricting low altitude cloud base, small caliber fire hit one of the Ouragans. The pilot was wounded. One fragment struck his eye. With vision in one eye only, he still managed to return to his base, helped by his section leader. In addition to the effort of the interceptors and attack aircraft, the IDFAF also operated its helicopters for CASEVAC and light planes for artillery spotting. Even more significant than the operation of its jet fighters, the operation of the S-58s and Drors marked the close cooperation with the ground forces, so advocated by the IDFAF, to become such a distinct feature of the Israeli retaliation actions and special operations in the following years.

The Israeli action in Samoa resulted in a situation of fluid unrest on the Egyptian border, possibly due to Arab solidarity. The opportunity was explored by 101 Squadron to raise its score to three kills with Mirages on November 29, when a pair of Mirages were scrambled from Hatzor air base to intercept Egyptian fighters which were considered a danger to an IDFAF observation lightplane at Nitzana. The pair arrived at the area to discover two EAF MiG-19s. Each of the Israeli Mirages launched its single R530 SARH AAM at the Egyptian section. One of the Matras, fired by the Israeli leader Seren Michael in Mirage number 52, found its target and the Egyptian fighter became the first victim of the new weapon in Middle East air combat history. After achieving a 50% kill ratio with their AAMs, not a particularly bad result for those days, the Israelis closed in and the leader shot down the second MiG with his cannon in the classic action — first firing AAMs from as long a range as possible and then closing in to finish off the remaining opponents with the cannon. Interestingly, that R530 maiden kill was the only one achieved by that weapon in Israeli service.

Thus, as 1966 drew to its end, the IDFAF, under the command of an officer during a trial period, was already credited with five kills. All the kills were achieved by the Mirages and, quite significantly, these were shared by all three units operating the type.

Also during those eight months the IDFAF flew many other operational sorties — attack, recce, CASEVAC, observation, etc. — and obtained the priceless MiG-21 from Iraq. It was a period of achievements, but the IDF General Staff was not yet impressed and Hod's trial period continued until a really solid achievement which no one could ignore was obtained in April 1967, barely less than a year since Hod entered his new office.

During the morning of April 7, 1967, a skirmish developed on the Syrian border . . . which was almost routine at the time. Unusual was the fact that this time the local incident, southwest of the Sea of Galilee, developed into a full scale artillery duel all along the Syrian border by noon. The Syrians not only enjoyed numerical and technical superiority as far as artillery was concerned but also enjoyed the benefit of the topography, shelling the Israelis down below while their artillery spotters having a plain view of all Israeli targets.

Thus the IDFAF was called upon to attack the Syrian artillery batteries at the Golan Heights and, as was the custom of the day, the attacking aircraft enjoyed the confidence that the Mirages were on CAP. By roughly 14:00 hours the first air combat of the day developed when Seren Iftah Spector led Seren Benjamin Romach, both of 101 Squadron, to intercept a quartet of SAF MiG-21s over the Golan Heights. They chased the Syrians all the way up to Damascus. Spector scored the first of his eventual tally of fifteen confirmed kills in that battle while Romach scored his only kill as, just three months later, he lost his life during the Six Day War. The two kills raised the total scored by 101 Squadron as a Mirage unit to five and it seems that the other Mirage units made an effort to close the gap. At about 16:00 hours, a pair from 117 Squadron intercepted two Syrian MiG-21s over the northern Golan Heights and in the long battle that developed, the Israeli leader, Seren Ezra Dotan, scored his first personal victory and the third Israeli victory of the day.

But the big day was not over yet as only a quarter of an hour later, and at the opposite end of the Golan Heights more than 50 km to the south over the Syrian-Jordanian border, Mirages of both 117 Squadron and 119 Squadron intercepted a quartet of SAF MiG-21s and shot down three of them. The three Fishbeds crashed behind the Jordanian border with Rav Seren Peker scoring the second of his seven kills and Seren Avner Siepak his first. But it was Sren Avi Lanir of 117 Squadron that captured the headlines when he passed, with his Mirage number 60, through the fireball of the MiG-21 that he shot down, emerging in an all black Mirage and virtually seeing nothing through the cockpit canopy. Seren Moti Yeshoron, the vicecommander of 119 Squadron, immediately joined the black Mirage to cover Lanir's egress from the battle area and to direct him to Ramat David air base where Lanir landed safely to celebrate his first victory and a unique achievement indeed.

That was it. In less that three hours, the three Mirage squadrons shot down one enemy fighter more than they did during the preceding nine months and, with eleven confirmed kills to its credit, it was clear which fighter was the new ruler of the Middle East skies. Not only was the air-to-air action of April 7 so successful but also the air-to-ground action silenced the Syrian artillery. By the day's end it was clear that the trial period was over. Moti Hod was promoted to Aloof and the stage was set for the first great concert by the IDFAF, the Six Day War, in which the Mirage played as a virtuoso.

FLYING THE MIRAGE - Here is the testimony of the initial batch of Israeli pilots assigned to fly Mirages with 117 Squadron during the early years: ". . . . . . On the ground, the Mirage is like any other aircraft, but when you turn it loose, you know you are riding a thoroughbred. 0.9 Mach at 40,000 ft. Pushing the throttle forward to the limit I switch on the burner and am immediately kicked back into the seat as the engines boost to supersonic. The Mach counter starts to swing like crazy, 1.0, 1.1 . . . 1.3, 1.4 . . . 1.5, I close the burner and try to maintain speed. Then, another quick, as I use afterburner to climb further to 45,000 feet. Now I am set to take her to the limits. Pressing on with full burner, the Mach indicator reads Mach 2. But after a glance at the fuel gauge I throttle back. In the air you feel like an arrow-head, the delta wings are swept way aft, so from any angle your field of vision is unobstructed. From this altitude, you can see the whole country: 100 km to the north, less to the hazy south. The sea lies on the west, and the Jordan valley and sea of Galilee to the east. I descend to 35,000 feet, the fuel is low and I must prepare to land. This engine swallows 400 liters a minute when on the burner, and in the ten minutes from take-off, it has used 2,000 liters. At that price you can't afford to take her supersonic just for fun.

Final approach. Gear down, I am turning into the final leg at 240 knots. As the speed reduces, the nose lifts and you can't see much. I put the aircraft down by looking at the runway's sidelines. At 175 knots and seven feet over the runway, I throttle back to idle and pull back the stick to break the flare. The aircraft touches down gently and the smooth magic is broken by the earthly bumps of the runways blocks and patches. I release the chute, slow down and taxi to the recovery point at the entrance of the underground shelter. Flying a Mach arrow, this aircraft met every expectation that I had and much more . . . ."

3

The Six-Day War

Although the actual date in which the Six-Day War broke out was June 5, 1967, it is worthless to describe the events beginning with that day without an extensive overview. This conflict was not a detached event in the modern history of the Middle East. It was merely another round in the Israeli struggle to survive an unfriendly environment. Surrounded by hostile Arab neighbors and the Mediterranean Sea, Israel, with its tiny population of only three million at the time, had become the prime target of the Arab world since its creation back in 1948. Moreover, it was the only uniting force in that world of conflicting interests and historical disputes and it was this fact that led to the immediate breakout of the Six-Day War.

Since its defeat in 1956, Egypt kept its border with Israel relatively quiet and the burden to continue the armed struggle against the Jewish state fell on the shoulders of Jordan and Syria. As a consequence, both nations suffered severely from Israel's retaliations while Egypt's President Nasser, the claimed leader of the Arab world, expressed his verbal solidarity as a compensation each time Israel retaliated. Surely this was not enough as far as Jordan and Syrian were concerned.

The first time that Nasser really acted was in early 1960 when the Egyptian Army entered the Sinai Peninsula following an Israeli retaliation operation against Syria. That incident had no immediate implications since Nasser withdrew his forces from Sinai after three weeks of tension. However, it was a good lesson to Israel and it became apparent that another round in the Arab-Israeli struggle was indeed only a question of timing rather than a remote possibility. For the Syrians, the Egyptian action was not radical enough; but when, during the skirmishes of 1964, Egypt did not even move its army into its forward positions, the Syrian disappointment with Egypt was quite plain and Nasser felt that his position as the claimed leader of the Arab world was threatened. By the autumn of 1966, following Operation Magresa, Jordan too blamed Egypt in the desertion of the Arab-Israeli battlefield.

That month Egypt and Syria signed a mutual defence treaty. So, when Israel and Syria clashed again on April 7, 1967, Nasser had a legal, rather than just a moral, obligation to support Syria. The Syrian accusations that Egypt did not come to its aid were enhanced by the Soviets. They informed the Arabs that Israel concentrated several brigades facing the Golan Heights to attack Syria, and thus Nasser felt that he should act to deter Israel from attacking Syria. After all, the mutual defence treaty was signed only six months earlier. The Soviet information was false, only God and the Kremlin knows why they distributed such intelligence, but by May 14, 1967 Nasser decided to order his army to enter Sinai, occupied only by UN forces since the Israeli withdrawal ten years earlier.

From that point the snow ball accelerated without anything to stop it. On May 20, Egyptian paratroopers were air dropped at Sharm el-Sheikh and Israeli shipping through the Tiran Strait, the formal Casus Belli of the 1956 Sinai Campaign, was stopped. Two days later, on May 22, Nasser visited an Su-7 squadron at Meliz air base at Bir Gafgafa, Sinai, and declared that if the Israelis wanted a war then they were welcomed. Israel, with Operation Rotem (Brrom Bush) of 1960 still fresh in mind, mobilized its reserves. A period of three weeks, called the Waiting Period, commenced. During these three weeks the diplomatic effort to solve the conflict resulted in a dead end. Simultaneously the armed forces of both sides were refining the operational plans, training the forces and making an effort to obtain as much information as possible about the deployment and intention of the other side.

One of the basic tools used to obtain such intelligence is aerial photography and during the Waiting Period both sides made an effort in that direction. The Egyptian photographic platforms were MiG-21Rs and these were used to photograph the IDF forces in the Negev. The technique was quite simple: the recce MiG flew to Jordan and then turned west and crossed the Negev through Dimona and Nitzana at high altitude and high speed, using afterburner, back home to Sinai. Those flights were codenamed Ephroakh (Chicken) and the IDFAF Mirages were scrambled to intercept them but without any apparent results as time and space were just not right to allow the Mirages to climb and accelerate quick enough to catch the MiGs before they left the Israeli airspace back to Sinai. It may be that a SEPR 844 boost rocket motor could have solved the problem but the IDFAF did not procure that item and thus four Mirages were detached to Hatzerim air base in the Negev. That move did not solved the problem either, the Mirage climb rate was just not good enough to nail the intruders. So it was Sgan Aloof Benjamin Peled, C.O. Wing 4, that suggested to intercept the reconnaissance MiGs with the ex-Iraqi MiG-21F-13 number 007 which had superior climb performance.

The suggestion was accepted and the MiG was armed with two RAFAEL Shafrir 2 AAMs, still under development. It was hoped that the MiG-21R, flying at the cold stratosphere with full afterburner, would be an ideal target. The MiG pilot, Sgan Aloof Dani Shapira, was afraid of innocent mistakes. Shining red stripes, to

A pair of Mirage IIICJs —probably from 101 Squadron— flying over rugged terrain in southern Israel, while on operational patrol. (Dassault)

assist identification, were added to the Israeli MiG and a pair of Mirages were used to escort the friendly MiG in the scramble. Alas, once the MiG-21F-13 was ready with its pair of Shafrir 2 AAMs, it was almost 5 June and the Israeli MiG versus Egyptian MiG air combat never took place. The last Egyptian incursion was recorded on June 4. Nevertheless, the availability of 007 caused Shapira to sit for four long hours in the tiny cockpit on the morning of June 5. In those hours the MiG-21 and four Mirages of each of the three squadrons were the only fighters available to protect Israel from Arab air attacks. The rest of the IDFAF combat aircraft inventory, plus some of its trainers, were busy in the execution of Operation Moked (Focus).

THE PLAN

Operation Moked was the codename of an emergency plan, of the sort to be found in the safes of every modern air arm around the world, which was completed in its basic form around 1963. The aim of the plan was to knock out the Egyptian Air Force — the largest, best equipped and most dangerous opponent — in a surprise attack. The plan was not improvised in the late spring of 1967. It was a well researched, planned and rehearsed operation which had its origins in the early years of the IDFAF. Destroying an air arm on the ground at its own air bases as the best option to gain total air superiority was not an Israeli invention. Guilio Douhet, the great proponent of air power, wrote in 1921 about such a possibility to obtain air superiority in his book, "The Command of the Air," while limited execution of that vision occurred in World War II.

As for the IDFAF, its first commanders were mainly ex-RAF servicemen and thus British oriented. They especially learned the doctrines used by the RAF in the Western Desert during World War II. The first operation of the IDFAF first fighters, the Czech Avia S 199s, was to be a surprise attack on El-Arish air base to wipe out the REAF's forces there. Although the attack had to be canceled due to other more urgent operational needs, the IDFAF retained that philosophy as one of its main guidelines in the event of war. During the early fifties the attack on enemy air bases was emphasized in the training and exercises of the IDFAF and operational orders were issued to the squadrons regarding an attack on the EAF's air bases in case of emergency. By 1956, through agreements with Great Britain and France, the IDFAF was not allowed to execute those plans and the EAF was wiped out by the Anglo-French forces. It took them several days to achieve that goal and quite a number of Egyptian aircraft survived.

The late fifties and early sixties saw the intensification of training regarding that subject. By the mid-1960s, the IDFAF Operations Department, headed by Sgan Aloof Rafi Harlev, who formerly commanded 119 Squadron, concluded the emergency plan that became known as Operation Moked. The man behind the plan was Seren, and later Rav-Seren, Rafi Savron, a navigator that headed the Attack Section of the Operations Department between 1962 and 1965. Another figure that played a dominant role in the creation of the plan was Rav-Seren Yak Nevo, who assumed a position within the Operations Department in 1961 after leaving the command of 105 Squadron. This plan was a masterpiece although quite a number of points, such as the exact H-Hour, were left open to be decided just before execution by the decision makers. The main attributes of the plan were:

GOALS

To neutralize the enemy air forces through the damage of its air bases' runways and the destruction of its aircraft on the ground.

METHOD OF ATTACK

Since several air bases were to be attacked simultaneously, each by several formations one after the other, it became plain that only small formations of four aircraft could be spared for each attack. Thus it was decided to ignore the air defence suppression mission and concentrate on the two main targets: the runways and the aircraft. Each quartet was to make a single bombing run to crater the runways (the Vatours sometimes made two such bombing passes because of their higher bomb load of eight 250kg bombs carried externally on MERs with additional fuel tanks in the weapons bay) and then return for several strafing passes to destroy the parked aircraft. The EAF still had no Hardened Aircraft Shelters (HAS). The number of strafing passes was calculated according to the predicted ammunition and fuel state and the estimated time on target since each formation was to be followed by another in order to keep the defenses under pressure and maximize the damage. Usually the number of strafing passes was to be two or three but the final decision was left to the leader of the formation, who also had to consider the AA defenses, and thus at least one formation made as many as five strafing passes.

During the attack the pilots were to ignore the AA fire and concentrate on the higher value targets. Priority was given to the destruction of the Tu-16 bombers and the MiG-21 interceptors.

It must be stressed at this point that not all runways were to be cratered. The plan was for the 55 Paratroopers Brigade to capture that air base in an airborne raid which was canceled because of the success of Operation Moked and the faster than anticipated follow-up advance of the IDF Armoured Forces.

LOSSES

It was predicted that the IDFAF might lose as much as 25% of the attacking force, roughly 45 aircraft. Losses were expected either to the defenses, AA fire and interceptors, or to accidents sustained during the intensive operation and ultra low level penetration.

The losses to be inflicted on the enemy, the EAF, were estimated at 80% of its combat inventory, something like 190 to 200 aircraft. It is interesting to note that the EAF considered an Israeli preemptive attack as a likely possibility, estimating that in such a scenario the IDFAF might destroy about 20% of its combat inventory, roughly 50 aircraft. As always, the actual figure was between the two estimations but much close to the Israeli view than to the rather optimistic Egyptian concept.

ROUTES

Every effort was made to conceal the incoming aircraft and one of the primary tools of the period to avoid radar detection was an ultra low level flight, less than 100 feet AGL. Moreover, the Jordanian radar network was able to observe every IDAF aircraft taking off and climbing above tree top level. Thus, the formations could not assemble at a relatively safe altitude and then dive to low level but had to keep themselves as low as possible immediately after take-off. This affected safety and made the assembly of the formations a demanding task. It also meant that all the ingress part of the mission profile had to be flown at ultra low level, restricting the radius of action as fuel consumption at low altitude is considerably higher than at medium or high altitude.

Three main streams were planned. The aircraft to attack the EAF air bases in the Nile Delta and the Suez Canal Zone were to fly west over the Mediterranean Sea. The aircraft to attack the forward EAF air bases in Sinai were to fly south to the Negev and then turn west to Sinai. Finally, the aircraft to attack the distant EAF air bases at Upper Egypt were to fly south over the Negev, Aqaba Gulf, and the Red Sea and then turn west to Upper Egypt.

Following the loss of surprise, after the first formation attacked, the incoming formation flew in a straight line and at high altitude. This allowed, as an example, the second Vatour formation to attack Beni Suef air base, and the third, to take-off from Ramat-David air base rather than Tel-Nof air base as the first formation.

WEAPONS

Later, after Operation Moked was executed, it was claimed by the popular press that the IDFAF used "secret weapons" to damage the Egyptian air bases and destroy its aircraft. This was not true. The only weapon specifically developed to attack air bases was the Runway Piercing Bomb (RPB) which was developed by IMI (Israel Military Industries) in cooperation with Matra of France. The RPB was a bomb equipped with a parachute and a rocket. After released at low level and high speed, the parachute was deployed to slow down the bomb and point its nose down. This served two purposes: the safety of the attacking aircraft was enhanced while the ballistics of the bomb were improved, giving it a better chance to inflict more damage. The rocket was fired and the bomb, nose down, was accelerated toward the runway to improve its penetration. Finally, a long delay fuse ensured that it would explode well after penetrating the runway, causing a large and deep crater. Those bombs were used only against the more important air bases, such as Cairo West and Kabrit, because their stock when the Six Day War broke out was still limited: a total of 253 bombs of two sub-types were available of which 169 were dropped during the first day of the war. The RPB had one major deficiency — its drag was considerably higher than a regular iron bomb, a fact that decreased the radius of action of the carrying aircraft. Other weapons used were iron bombs, mostly 250kg, with suitable fuzes to inflict the required damage and 20 or 30mm aircraft cannons used for strafing attacks.

DECEPTION & ELECTRONIC WARFARE

An elementary part of the plan was the deception of the enemy. A part of that deception effort was to be the simulation of a regular training day by the Magister trainers of 12 Flying School, including all the relevant radio traffic.

In parallel, an effort was to be made to jam the radars and communications of the enemy. It is known that an active/passive EW was developed by RAFAEL and employed operationally during the Six Day War and that Israeli intelligence made a special effort to locate the positions of radars and command and communication centers. It is also reasonable to supposes that ELINT was also gathered before the war broke out. In addition, the armed Magisters of 12 Flying School were tasked to destroy four front-line Egyptian radar stations in Sinai in parallel to the attack on the EAF air bases.

BRIEFINGS

The importance of a highly detailed briefing can not be underestimated. Every Israeli pilot participating in Operation Moked knew all the relative information about his assigned target, including aerial photos, resident units, route planning, etc. The pilots were ordered to persist with their mission at

any cost, even if an aircraft in the formation crashed enroute to the target or any unpredicted problems popped up around the target area. Under no circumstances were the pilots allowed to break radio silence.

The final briefings to the pilots were to take place immediately before take-off. The briefings at squadron level commenced at about 05:00 and take-off was from 07:00 onwards. The air base commanders and air wing commanders were briefed on the afternoon of the previous day, June 4, while the squadron commanders learned of the decision to execute Operation Moked when they were ordered to open sealed envelopes that evening. Thus the secrecy of the operation was preserved for as long as possible.

TARGET ALLOCATION

Two main factors dictated the allocation of targets: the radius of action of the type concerned and the known resident units at the target under question. Since all the ingress part of the planned mission profile was at low level, the radius of action of all types was decreased considerably. Thus the types with the shortest radius of action, the Ouragans and Mystères, were assigned the targets in the Sinai Peninsula and the Suez Canal Zone. The longer ranged types, the SMBD, the Vatours, and the Mirages, were assigned targets in the Suez Canal Zone and the Nile Delta, while targets in Upper Egypt were allocated to the Vatours.

However, it was clear that some of the aircraft based at Ramat David air base had to be deployed to more southern bases to shorten the involved range. All Ouragans based at Ramat David were thus deployed to Lod air base while the Mystère and Vatour formations assigned to attack the more distant targets, Kabrit air base at the Suez Canal Zone in the case of the Mystère and Beni Suef in Upper Egypt in the case of the Vatour, were deployed to Tel Nof.

The second consideration was the type of fighters to be expected at the targets and an effort was made to allocate targets that did not contain MiG-21 fighters to the Ouragan and Mystères. This, however, was a secondary consideration and indeed the Ouragans that attacked Meliz air base at Bir Gafgafa, Sinai, had to face the defending MiG-21s as did the aircraft that attacked Abue Sueir air base several dozen kilometers to the west. That problem was partially resolved by the allocation of at least one formation of Mirages to air combat over the Suez Canal Zone upon its return from its assigned target west of that area, naturally at the expense of several strafing passes to conserve fuel and ammunition for the anticipated air combat over hostile air bases.

AIR DEFENSE

While the Israeli air armada was to attack the enemy air bases, Israel was entirely, or almost entirely, defenseless against enemy air raids. It was obvious that a number of interceptors were to be left behind on QRA (Quick Readiness Alert) but the actual number caused a severe headache to the planners. A small number might prove inadequate in the event of an enemy retaliation during the turnaround of the aircraft that were to return from the first wave. Too many and the number of aircraft to actually inflict damage might prove insufficient to neutralize the enemy air forces. The view that was accepted was that of Allof Hod, who advocated the first option and each of the three Mirage units was to contribute a quartet of interceptors on QRA to make a total of only twelve. Additionally the MiG-21, number 007, manned by IDFAF CTP Sgan Aloof Shapira, which was readied for the interception of intruding EAF reconnaissance MiG-21Rs, was to be kept on QRA at Hatzor air base. Backing the small interceptor force were the Hawk SAM batteries and the 40mm Oerlikon L70 AA guns which were dispersed to protect key installations.

DAMAGE ASSESSMENT

The problematic nature of reports from the fighting units can always lead the decision makers to false conclusions. The most common problem is that of overestimating enemy losses. This usually results from multiple counting, for example, when more than one pilot attacks the same target, when optimistic reports are given, or when dummy targets are attacked. So, in addition to the reports of the pilot, damage assessment must rely upon reconnaissance flights and, indeed, in photos of the EAF air bases, released by the IDFAF after the attack, the shadows of the photographing Mirages and Vatours can be clearly observed.

It is highly possible that separate reconnaissance missions were despatched but at least on one occasion the lead aircraft of an attacking formation was a reconnaissance aircraft. This was the third formation to attack Beni Suef air base in Upper Egypt. The lead aircraft was Vatour IIBR number 33 (nicknamed Ha'akh Ha'gadall — The Big Brother —) which was flown by Seren (reserve) Gideon Magen with the navigator Raz. They bombed the air base with their eight 250kg bombs and then, unlike the rest of the Vatours which strafed the air base, photographed Beni Suef.

INTELLIGENCE

Operation Moked depended heavily upon precise intelligence. Every detail, as meaningless as it might seem to be, was carefully studied by the IDFAF planners. The readiness of the enemy interception pilots, its CAP and interception procedures, the thickness of the runways' concrete, the daily routine of its units and more, much more.

The exact location of the enemy's units was naturally a highly important factor. The movements of the enemy's air units were closely monitored and, in fact, changes were made each time such a redeployment was discovered. An example is the evacuation of the MiG-17s from Jabel Libni air base in Sinai to Kabrit air base in the Suez Canal Zone just days before 5 June. As a result, the Mystère formation which was to attack Libni had to switch to Kabrit, but, due to the longer flight involved, had to take-off from

Although heavily censored, this photograph of a 101 Sqn Mirage IIICJ is interesting as this aircraft has been retrofitted with a Thomson-CSF BU radar warning receiver on the fin trailing edge, just above the rudder. Later this equipment would be replaced by Israeli made kit. (IDF)

Tel-Nof air base instead of Ramat David, a fact that complicated planning.

NAVIGATION

These were the days (in the IDFAF) before INS and GPS and formation leaders had to navigate with the compass, a watch and a map. This fact put a lot of stress on the leaders. Moreover, the navigation was mostly over the sea, the desert, or the Nile Delta and in all three areas distinct landmark is just nonexistent. In addition, the leaders had to stick exactly to the preplanned timetable which complicated things even more. Nevertheless, the leaders proved themselves quite well and only a few navigation errors took place. An unexpected problem encountered by quite a number of leaders was the morning mists which made the final identification of the target quite problematic. In that context it can be said that the EAF air bases at the Nile Delta were harder to locate than those in the desert because of the surrounding terrain. The problems concerned with navigation and target acquisition disappeared after the first formations attacked the air bases as the columns of black smoke were visible for dozens of kilometers.

SAFETY

It can not be said that safety was set aside but it was clear that when safety and operational needs conflicted, it was the later consideration that mattered. The air crews were briefed to keep the taxiways and runways clear from any obstruction at any cost. That meant that an aircraft with a technical malfunction had to be pushed aside and that a pilot experiencing a technical failure during take-off had to crash his aircraft deliberately anywhere except the runway. However, when a technical problem was discovered before take-off, no chances were taken and that particular aircraft was not dispatched; as a result, several formations had to continue as three-ship flights instead of a quartet. Enroute, and before H-Hour, radio silence had to be preserved even if a pilot ejected, crashed into the sea, landed, or aborted his mission. Moreover, the leaders were instructed to continue their mission even if such an incident occurred. SAR helicopters were naturally on alert but without radio were unable even to know of an emergency had occurred, let alone the exact location of the pilot in distress. This had to be dealt with only after H-Hour.

SCHEDULING

The planned schedules left little to improvise as they were very tight. The leaders had to stick exactly to the timetables from the wakening of the aircrews, which commenced at 03:30, to the attack. Because so many aircraft had to take off from so few air bases, at such a short time, the pilots had to stick to their exact timing and perform all the routine checks within the permitted time as a delay in take-off would have affected the whole plan. Exactly the same role applied to take-off, which had to be smooth and quick, to the assemblage of the formations, to the correct and timely navigation and to the actual attack.

SECOND WAVE

Unlike the first wave, which was almost over planned, in the second wave targets were to be allocated according to the development of the battle, hence the importance of damage assessment. The second wave, which commenced almost two hours after H-Hour, witnessed a lot of mission changes, although not as many as the third wave. Pilots were briefed about a target in the briefing room only to be caught in the cockpits by a messenger with other maps and new orders. Some formations even changed three targets before finally leaving the ground. This was mainly attributed to the success of the first wave and the uncertainty to the reaction of the other Arab nations.

Thus, in order to summarize, Operation Moked called for a coordinated attack on the main air bases of the EAF. Each air base was to be engaged by a formation of four aircraft which were to bomb the runways, to make sure that the air base would be inoperable for the coming hours, and then make several strafing attacks to destroy parked aircraft as the EAF had not any HAS yet. The number of strafing runs were restricted by the estimated ammunition and fuel states but the leader of the formation had the freedom to abandon the attack, if the AA fire was to prove too dangerous for example. Nevertheless the leaders knew that it was highly important to make as many strafing attacks as possible since each formation was followed by another one, at something between five and fifteen minute intervals, to keep the enemy defenses under pressure and ensure the maximum damage. The targets were allocated to the types of aircraft mainly as a consideration of the radius of action while the highest priority, as far as the target were concerned, was given to the MiG-21 interceptors and Tu-16 bombers.

Many points in the plan were left to the decision makers. The exact timing to commence the attack was one such point, the number of interceptors left behind on alert was another. Even targets for the second wave were not prematurely allocated but were to be designated to the units according to the developments of the battle. The plan was consistently updated as the EAF moved its units from base to base and as new intelligence information arrived. The squadrons were kept briefed about the developments and each squadron knew its primary and alternative targets. However, above all stood years of preparation, planning and training.

THE FIRST WAVE

On June 2, the IDF CoS Rav-Aloof Itshak Rabin visited 119 Squadron at Tel-Nof. Rabin was impressed with Rav-Seren Ran Peker's bunch of professionals and it is claimed that this visit made Rabin believe that Operation Moked might well be executed as planned. Anyway, it was the Israeli government decision to end the Waiting Period with a preemptive attack instead of waiting until the Arabs would be prepared and attack first. But, above all, air superiority had to be achieved quickly.

On June 4, the day before D-Day, a critical decision was made: H-Hour would be at 07:45. This was an odd hour to choose since traditionally air arms attack at dawn or dusk, but the IDAF planners rejected those two options since the former was just too obvious and the later gave no time to follow an initial success with a second wave (as in those days night operations could not give the results of daytime attacks). So 07:45 hours was chosen because it left almost a full day to complete the mission and to support the IDF ground forces that were to launch their attack at Sinai simultaneously. Moreover, it was claimed that the EAF was waiting for a dawn attack and thus keeping its CAPs up at dawn but those were already back on the ground by 07:45 with the feeling that another day passed without any operational activity. This was also thought to be the way of thinking of the Egyptian planners and decision makers that were to be on their way to their offices at 07:45, thus losing several critical moments. Not less important was the fact that H-Hour at 07:45 allowed the IDFAF aircrews to sleep well, or at least more than in the case of a dawn attack, and thus it was felt that more could be squeezed out of them during the first day of operations. Finally a strike at 07:45 meant that the navigation would be at daylight while it was hoped that the morning mists would be evaporated by that hour by the shining sun.

So it was to be 07:45 and during the afternoon hours of June 4, the commanders of the air bases and wings were briefed while later on, in the evening, the commanders of the squadrons were also informed. In parallel, the last minute arrangements had to be made such as the deployment of Vatour and Mystère formations from Ramat-David to Tel-Nof. These were ferried in singles and at low-level, sneaking in to Tel-Nof air base as one pilot described it. At least one of the Mystère pilots guessed correctly what all that activity was about while another clue was given in the evening when the aircrews were ordered to get a good sleep, from about 21:00 hours, as tomorrow would be an exhausting day.

Next morning, before dawn, at 03:45, the wakening up of the aircrews, beginning with the commanders and leaders, commenced. The briefings at all bases were at squadron level and commenced at 05:00 varying in the personal style of the briefer rather than in the content of the brief. The main issues described in those briefings were already described and, anyway, the briefings were only to refresh the memory of the pilots and navigators since they knew their targets and the general plan so well. Next, around 06:30, the aircrews left the briefing rooms for their aircraft, which were double checked. No chances were taken and aircraft which developed malfunctions were not to take-off.

Silently, with only the "routine" radio traffic of the 12 Flying School Magisters on deception in the air, the quartets assembled and taxied to the runways. Most of the aircrews surely feared, as we all do, from the unknown but the morale was high as these men knew that they were chosen to execute a plan that they practiced for years — a rare opportunity indeed. Moreover, they felt that the destiny of Israel was in their hands, the hands of the professionals whose skills were sharpened during the three weeks of intensive training that preceded June 5.

Finally at about 07:00 the first formation took off, followed at short intervals by all other formations. An aerial armada was flying at ultra low level and in total radio silence to write a glorious chapter in the history of air warfare. But in the tiny cockpits of the Israeli combat aircraft, the stress was almost unbearable. Almost scratching the deck, holding desperately to a precise navigation over unfamiliar terrain, unable to communicate with each other and totally devoid of the complete picture, the pilots and navigators simply pressed on. Some of the leaders were quite not sure about their navigation but stuck to their compasses and maps so when the time came, they zoomed up and lowered their wings to locate their targets and then dive to bomb the runways. Most of the leaders were surprised to encounter morning mists or low clouds that made target identification quite difficult, but since in general the navigation was quite accurate, this problem was overcome by determination. Of all the formations of the first wave only a single formation made a major navigational error, leading to the attack of eleven air bases instead of the ten originally planned for that first wave. Now follows the attack results on those air bases during the first wave.

ABU SUEIR (AB 229 EAF code)

This air base was a high value target since it housed the 61 Air Brigade with at least 21 operational Il-28 bombers that had offensive plans for attacking IDFAF air bases. Also, this airfield housed a MiG-21 interception unit which posed quite a threat to the attacking formations.

The first formation to attack was led by Rav-Seren Amihai Shmueli, C.O. of 117 Squadron. They took off from Ramat-David

Highly polished Mirage IIICJ number 259 of 101 Squadron while on an open day at Hatzerim air base before the Six Day War. Note the 1.300-liter drop tanks in the foreground as well as the DEFA 30mm cannon. (author collection).

at 07:15 hours and attacked at 07:45. Navigation was achieved by compass and watch as no apparent landmarks were available, but the EAF airfield was clearly observed through holes in the cloud layer and after bombing the runways, one after the other, the four Mirages strafed the parked Il-28s. It seems, however, that the damaged runways were used by the MiG-21s to scramble and later formations met fierce resistance. These consisted of Mirages, 101 Squadron also participated, and the Vatours of 110 Squadron. A formation from that unit led by Seren Shlomo Keren that attacked at 08:40 hours was intercepted by four MiG-21s, which the Vatour number three (element leader), Seren Ben-Zion Zohar, engaged alone to allow the other Vatours to press on their attack, later on bombing the bases himself and returning to Ramat David on one engine due to fuel starvation. The Vatour might be claimed to be a bomber, but according to a pilot that flew them with 110 Squadron for about ten years, it was a very good dogfighter, being overstressed in training quite frequently. Nevertheless the MiGs at Abu Sueir did claim one of the Vatours, Segen Dan Manor ejected west of the Suez Canal and an effort to rescue him by a Super Frelon in the coming evening failed. He was take POW the next day.

Once it was clear that the first wave was a success and no significant retaliation could be expected, a pair of Mirages of 119 Squadron, of the four on QRA from that unit, were scrambled just before 09:00 hours to Abue Sueir to intercept the bothering MiG-21s. Very low on fuel, since they were configured for interception over Israel, the pair had to engage quickly and this they did with enormous success. The wingman, Segen Giora Rom, shot down two MiG-21s and just barely managed to return to Tel-Nof. His leader, Seren Eitan Carmi, was not so fortunate and although also shooting down two MiG-21s, he had to eject from his Mirage due to fuel starvation. The successful action of the two Mirage pilots, both to become aces with Rom attaining the status of the first Israeli jet ace, while Carmi eventually raised his score to eight kills plus an AS-5 missile, ended the interference of MiG-21s over Abu Sueir, but that was to be the advantage of the second wave as the first drew to its end at around 09:00.

All in all, Abu Sueir was attacked by ten formation during the first wave; while the total number of bombs dropped was 102, this figure including quite a number of RPBs. In spite of the fierce resistance experienced, the goal was achieved with most of the Il-28s destroyed on the ground, four Fishbeds shot down and the loss of two IDFAF combat aircraft.

BENI SUEF AB

This was the only Egyptian air base in Upper Egypt to be attacked during the first wave. Because of its location, the Egyptians figured that it was beyond the reach of the IDFAF and hence located part of its Tu-16 force there which had the order to retaliate if Israeli aircraft attacked EAF air bases. In any event, this was just not possible since the Tu-16s were destroyed on the ground.

The first formation to attack consisted of three Vatours, the fourth had a technical malfunction before take-off, which had launched from Tel-Nof, since the necessary range at low altitude was just impossible if Ramat-David had been used as the launch point. The three Vatours flew at low level south, turned southwest at Ramon Crater in the Negev, then crossed Sinai and the Gulf of Suez and entered Egypt. The formation encountered morning mists but managed to locate the target and made two bombing runs before several strafing passes were performed to destroy the parked Tu-16s. No interceptors were encountered and the flak was light and not effective.

The second Vatour formation to attack was led by Rav-Seren Levy Zur, 110 Squadron C.O., in a Vatour IIN with the navigator Seren Ori Talmore. That formation took off at 07:25, reached the target area with the aid of smoke columns and destroyed the remaining Tu-16s after the usual 110 Squadron pattern of two bombing runs. The pilots of that formation reported that SA-2 missiles were launched at them but were not a threat to the Vatours that flew at very low level.

The third formation to attack was a quartet of Mirages from 101 Squadron, but they found that the preceding formations did most of the required work there and thus bombed and strafed before leaving the target area to the third Vatour formation from 110 Squadron that, this time took-off from Ramat-David, and simply flew in a straight line and at high altitude as the element of surprise had vanished by that time. The leader was Seren (reserve) Gideon Magen who flew in Vatour IIBR number 33 with a navigator. Each of the aircraft carried auxiliary fuel tanks internally in the bomb bay and eight 250kg bombs on MERs under the wings. Before reaching the target area they dropped to low altitude and they too encountered morning mists but the columns of black smoke easily identified the location of their target. Although none of the Tu-16s remained in a worthy condition for a strafing pass, the Vatours made two such runs during which they also turned their attention to the control tower and other installations. Naturally the Vatour IIBR had no guns and it was used to photograph the air base.

The raids on Beni Suef were a complete success. The Egyptians were caught totally unprepared as they thought that this base was beyond the reach of the IDFAF, and thus suffered heavy losses with all worthy targets being destroyed. The attackers suffered no casualties and only one aircraft, a Vatour from the first formation flown by Segen Hertzel Bodigner, was slightly damaged by the local flak, but its pilot returned home without any apparent difficulties.

MELIZ-BIR GAFGAFA (AB244, later REFIDIM AB by the IDFAF)

Meliz air base at Bir Gafgafa in Sinai was the major EAF installation in that peninsula. It housed several combat units, including No. 20 Squadron with MiG-19s, with operational orders to attack Hatzor air base in case of war. It is certain that a MiG-21 unit was also resident, while the Su-7 OTU was the unit which President Nasser visited at Meliz air base on May 22, although the unit was not yet operational and by June 5 its aircrews were assigned to 31 Squadron with MiG-17Fs at Kabrit. In addition, Israeli pilots that attacked the airfield reported about the destruction of MiG-17s. This may not have been true and might be related to misidentification or the attack of dummies or decoys. Finally Meliz contained a command post from which operations could have been controlled, usually assisted by an Il-14 flying as a communications relay.

Four Mirages flying over northern Israel. The first generation Mirage was catapulted to fame and its name entered into aviation's hall of fame tanks to the exploits of the original three IDFAF squadrons (101, 117 and 119) equipped with the type. (IDF)

Meliz was the principal target of 113 Squadron and three Ouragans of the Wasp Squadron were lost in roughly fifteen minutes to the heavy flak. The first quartet attacked at 07:45 and was welcomed by heavy flak while the MiG-21s on QRA were scrambled. Two managed to take-off just as the runway was bombed but one of them was immediately bounced by an Israeli Ouragan. Although there is no doubt that the Ouragan was inferior to the MiG-21, such a situation was one of the risks which the planners of Operation Moked accepted out of necessity. Anyway, the fact that the MiG took off just as the Ouragans dived to bomb and not earlier was a great advantage as the energy state of the MiG-21 immediately after take-off was quite low, while that of the Ouragans, after the bombing dive, was certainly higher. This advantage was used to good effect by Seren (reserve) David Yariv, who brought down the MiG-21 with cannon fire. However, the concentration of the Ouragan pilot in air combat cost him his life as immediately afterwards his Ouragan was struck by flak and he was killed. The second MiG left the area without trying to defend the air base, while the other three Ouragans continued their attack, claiming a total of six EAF MiG-21s and a single Mi-8 helicopter destroyed on the ground. One more Ouragan was lost to the flak, its pilot Seren (reserve) Mordecai Navon ejecting over the Mediterranean. He was picked up by an Egyptian boat to become a POW. Finally, the leader of that unfortunate formation, Seren Ran Alon, was also hit by AA fire, but he managed to make it back for a belly landing at Hatzor air base. Ten minutes after the first formation attacked, at 07:55 hours, the second Ouragan formation appeared at Meliz. It seems that the Egyptians were trying to launch their interceptors as three MiG-21s with the pilots in their cockpits were observed. All three were destroyed during the first strafing pass, although that formation also had a casualty to the intense AA fire. Rav-Seren Yoseph Salnet, 113 Squadron C.O. and the leader of that formation, managed to fly back until he was forced to eject over the Mediterranean barely five kilometers from the Israeli coast. He was rescued by an S-58 of 124 Squadron flown by Itshak Shanee and Itshak Shegev. The Egyptian airfield was also attacked by other formations, including at least one of Mirages and two of Mystères from 109 and 116 Squadrons; but, the basic work there must be attributed to the 113 Squadron that paid heavily with three Ouragans being lost. All-in-all, IDFAF pilots claimed the destruction of fourteen EAF aircraft on the ground at Meliz.

BIR TMADA (AIR BASE 260)

This was a relatively newly constructed air base and its existence became known to the IDFAF well after the plan of Operation Moked was finalized. As a direct result of that intelligence gathering, the plan had to be updated to include that airfield in the Sinai too. Bir Tmada was viewed by the EAF as a forward attack airfield and a such housed a MiG-17 unit, 25 Squadron of the 2 Air Brigade, as well as a detachment of the only operational Su-7 unit, 55 Squadron of the 1 Air Brigade. Because of its relatively close proximity to the Israeli air bases and the absence of MiG-21 interceptors, that air base was left to the Ouragans and Mystères, although a formation of Mirages from 101 Squadron also participated in the attack that commenced at 07:45 hours with a second formation attacking at 07:50 and a third at 08:00. The first formation to arrive consisted of 107 Squadron Ouragans that claimed eight MiGs, two Mi-4s, a single Mi-6 and a single Il-14. The second formation, 109 Squadron Mystères this time, claimed the destruction of the remaining four MiG-17s as well as Mi-4s and Il-14s. The formation of 101 Squadron observed a lone MiG-21 that crashed into the ground without any apparent cause. Thus the attack of Bir Tmada was successful with no losses and all targets were destroyed.

CAIRO INTERNATIONAL AIRPORT

This was not a planned target for the first wave as it was not supposed to contain combat aircraft that endangered Israel, although it certainly housed certain elements of the EAF. A single formation bound to Inchas — either Mirages or SMBDs — there are conflicting reports as to the type of aircraft involved, failed to locate its assigned target and instead attacked Cairo International as a target of opportunity. The pilots of this formation claimed that they observed combat aircraft on the ground. If that report was indeed reliable then it might have been that the EAF evacuated aircraft from air bases under attack in the hope that the IDFAF would not attack that installation.

CAIRO WEST AB

It was the home of No. 65 Air Brigade with Tu-16 bombers, including No. 95 Flight with Tu-16s equipped with AS-1 Kennel missiles. As such, Cairo West was a top priority target. Just like the other Tu-16 air base, Beni Suef, Cairo West was attacked by Mirages and Vatours although, contrary to the more distant target, the main burden fell on the shoulders of 101 and 119 Squadrons, rather than on 110 Squadron. Two reasons may be counted down. The first is that 110 Squadron main effort was Beni Suef and the second that the probability of hostile interception was higher in the Nile Delta than in Upper Egypt.

The first formation to attack Cairo West was composed of 110 Squadron Vatours that took off from Tel-Nof at 07:17 hours. This attack was followed by at least three formations of Mirages of which two were from 101 Squadrons and the third from 119 Squadron. The second formation to attack was led by Seren Oded

Marom of 101 Squadron from Hatzor. This formation also flew at low level, identifying the target area by the columns of black smoke and popping up to 9,000 feet for the bomb dive. Each Mirage released a pair of 500kg bombs and then three strafing runs were made during which the formation claimed five Tu-16s destroyed on the ground. It seems that the local flak was still weak at that stage — but the next formation, led by Seren Oded Sagee of 119 Squadron, was welcomed by a wild barrage of flak. The attack, at 08:00, was carried on as planned with the four Mirages popping up into a 35-degree, 450 knots dive, to enhance penetration of the two 7/14 second-delay fused 500kg bombs which were released at 2,500 feet. Then the quartet split for the coordinated strafing passes. Each pair came each time from a different direction with the leader attacking one half of the air base, while his wingman took charge of the remaining half to eliminate the risk of aerial collision between the Israeli fighters. The number four Mirage damaged a Tu-16 in his first pass, strafed a line of dummy MiG-17s on his second, damaged another Tu-16s on his third and only managed to hit an aircraft shelter wall on his last strafing run. The attack ended as the Mirages reached the deadline of 400 liters of fuel which was just enough to cover the distance of 300 to 350 kilometers back home to Tel-Nof.

The third Mirage formation to attack Cairo West was led by Rav-Seren Amos Lapidot, 101 Squadron C.O., carrying the first RPBs to be dropped on that air base. These weapons proved to produce considerably more drag than the regular 500kg GP bombs used by the previous formations. The leader was ordered to jettison the empty external fuel tanks in order to reduce drag and lower the fuel consumption but this measure was not wholly successful. On their return from Cairo West, the Atar engine of Lapidot flamed out over the El-Arish area and he glided to a successful emergency landing at Hatzerim air base. The other pilots managed to regain Hatzor although to both, numbers two and three, the powerplant flamed out immediately after landing while only 70 liters of fuel remained in the tanks of number four when he switched off the Atar. The formation claimed the destruction of eight Tu-16s, a single Il-28 and three MiG-21s.

So Cairo West and 65 Air Brigade were both out of action as the IDFAF first wave completed its job. No losses were sustained at Cairo West and the first formations were welcomed by weak flak which became more intense as the attack continued. The formations were not intercepted and although SA-2 batteries were located in the area, no formations to attack during the first wave reported about any launch although later on that day SA-2s were indeed fired at IDFAF aircraft in the vicinity of Cairo West AB.

The Egyptian Tu-16 strategic bombers were a priority target during the early stages of the Six-Day War, most were destroyed at Cairo West by Mirages of 101 and 119 Squadrons. (IDF)

the essential installations, such as the control tower, but only to destroy the MiG-17Fs and the AA guns. The mission was to be executed primarily by the Ramat-David-based Ouragan squadron, deployed at Lod for the occasion, with the aid of No 12 Flying School Magisters which were organized as an emergency unit commanded by Rav-Seren Aryeh Ben-Or, the C.O. of the Advanced Training Squadron, and staffed by elderly reserve pilots, a few QFIs and a small number of fresh graduates from 12 Flying School that did not finish their OTU when the crisis flared-up.

Two quartets of Ouragans commenced the attack on 07:45 hours with Rav-Seren Ya'acov Turner leading the first. The Ouragans, flying clean with only internal ammunition for the four 20mm cannon, caught the MiG-17Fs by surprise on the ground after these returned from an early morning sortie. Each Ouragan made three strafing passes and no flak or any other kind of opposition was reported by the Israeli pilots who claimed the destruction of six MiGs. The second Ouragan formation destroyed a seventh MiG-17 as well as an Il-14 which was observed in a hangar. Next to attack was a formation of armed Magisters led by Seren Arnon Livnat, the vice commander of the unit. Their target was the radar station in the vicinity of El-Aris and this was claimed to have been destroyed; but, on their return to Hatzerim, Livnat's Magister was hit and killed by AA near Rafah, the only IDFAF casualty in the attack on El-Arish. The original plan to capture the air base in an airborne assault was canceled when the success of the first wave was revealed and the understanding that the IDF ground forces advance in Sinai was to be swift and fast. 55 Parachute Brigade, a reserve unit actually, was instead deployed to the Jerusalem front.

EL-ARISH (259 AB)

The IDFAF was familiar with El-Arish since May 1948 when REAF Spitfires and Dakotas used that air base to attack Israel. Nineteen years later that same airfield hosted No. 18 Squadron of No. 2 Air Brigade equipped with MiG-17Fs and briefed to attack Tel-Nof in case of war. The air base was also heavily defended by 14.5mm AA guns and 37mm and 57mm AA cannons. Since it was the closest air base to the Israeli border, the IDF plan was to capture it by an airborne assault of 55 Parachute Brigade under the command of Mordehai "Motta" Gur who became the IDF COS in April 1974. Hence it was decided not to attack the runways and

FAYID (233 AB)

Fayid was a main Egyptian Air Force air base and in addition to an operations control center it hosted a MiG-21 unit, No. 40 Squadron of No. 5 Air Brigade, and the only EAF Su-7 unit, No. 55 Squadron of No. 1 Air Brigade. Another resident unit was a MiG-19 squadron. Thus although that airfield was within the reach of the Ouragans and Mystères, it also received the attention of at least one formation of Mirages. First to attack was a formation of three Mystères from 116 Squadron that were led by Seren Avihu Ben-Nun, the unit vice commander. The number three pilot had to abort the mission and made a successful emergency landing at Hatzerim. The formation, as almost all others before it, encountered low visibility conditions and a layer of stratus clouds but located the target and bombed the runways just as four EAF interceptors were about to take-off. Two of these were immediately set ablaze and in the following strafing passes that formation claimed the destruction of fourteen more EAF fighters.

An interesting event was that as the IDFAF fighters came out of the bombing run the leader observed an An-12 transport coming in to land at Fayid. He intended to order his formation to press on with the attack while he would shoot down the An-12 but then discovered that his transceiver was not able to transmit. He figured out that if he chased the Antonov without informing any of the other pilots, they would simply follow him and abandon the attack on the air base. Making the right decision, the attack on Fayid seemed more important than a kill, he ignored the An-12 that cleared the area with senior Egyptian Army officers on an inspection tour to Sinai.

In spite of excellent results achieved by the first formation, the Egyptian AA defenses soon overcame the initial shock and the following formations were welcomed by intensive flak which downed two Israeli fighters. Mordechai Pinto of 113 Squadron was killed when his Ouragan was hit by AA fire and Rav-Seren Yonatan Shahar, C.O. of 116 Squadron had to eject from his Mystère for similar reasons.

Shahar claimed to have destroyed a "Viscount" (possibly an Il-18) in his first strafing pass, a MiG-19 in his second and a MiG-21 in the third. The hit was either flak or debris from the exploding MiG-21 and Yonatan Sahar spent almost fourteen hours hiding from the Egyptian forces before he was retrieved under the cover of darkness by a Super Frelon that also tried to rescue Shlomo Manor but failed after evading two SAMs. Manor was taken POW the following day, 6 June. Other units that attacked Fayid were 109 and 117 Squadrons from Ramat-David. Thus Israeli losses on the attacks on Fayid amounted to one pilot killed and two fighters, but the air base was heavily damaged. In the coming days, a small number of EAF Su-7s operated in Sinai meaning that No. 55 Squadron was not completely wiped out. It is also possible that some of the MiG-19s and MiG-21s remained, but as a whole the first wave attack on Fayid can be considered a success.

INCHAS AB

This was, in all probability, the top priority target of the first wave as it was the home of an MiG-21 interception Air Brigade tasked with defending the Nile Delta. The elimination of that threat was to result in far fewer losses on the side of the attacking forces and many more on the side of the defenders.

First to attack Inchas was a formation of four Mirages of 119 Squadron led by the unit C.O. Rav-Seren Ran Peker. They were to take-off from Tel-Nof at 07:16 but a human error caused a five-minute delay which might have disrupted the timing of all the operation unless Peker flew faster than planned. Making up the lost time, they reached Pakuas, an Egyptian town which was the designated the pull-up point. Climbing to 6,000 feet, immediately followed by inaccurate flak, they settled for the 45-degree dive to enhance the penetration of the bombs that were released at 2,000 feet. The formation bombed the runway, witnessing parked MiG-21s with their pilots in the cockpits during that phase, and then turned to the strafing phase, making no less than five such passes in which they opened fire from a range of 400 to 1,200 meters, destroying many of the precious MiG-21s. The AA fire, at first weak, became more intense as the attack was pressed on but the Mirages were able to clear the area unharmed just as the next formation to attack, four SMBDs of 105 Squadron led by their C.O., Rav-Seren Aharon "Yalo" Shavit. Since the element of surprise was lost, the SMBDs were welcomed by a barrage of flak as they climbed for the bomb run and the number two pilot, Segen Dan Angel, received a direct hit, his Super Mystère exploding in the air even before the bombs were released. The three remaining jets carried on with the attack but again were unlucky when Shavit's SMBD was set ablaze by the AA fire. Shavit flew back home in the burning aircraft, regaining Hatzor where he made an emergency landing, saving the lives of both himself and the SMBD.

The third formation also consisted of four SMBD and their pilots witnessed some 12 to 15 burning MiG-21s. They added seven more and it seems that these three formations did most of the job, the only other detail required to be mentioned is that one of the formations bound to Inchas, either SMBDs or Mirages, was unable to locate its assigned target and attacked Cairo International instead, which was not intended to be attacked by the first wave. To summarize, the EAF elite interception unit was neutralized, and partly destroyed at the cost of a pilot killed in his SMBD, while a second jet was badly damaged, but repairable.

EL-SIR/JABEL LIBNI (248 AB)

This forward air base seemed to have been the least dangerous when compared with the other two, El-Arish and Bir Tmada. It was the home of a single MiG-15 unit, No. 24 Squadron of No. 12 Air Brigade, possibly the only remaining EAF operational MiG-15 squadron by that time. Hence, no threat of interception was foreseen and the airfield was attacked by IDFAF Ouragans which put the air base out of action without any losses or any significant events. Some 1,100 meters of the main runway's 2,800 meters remained intact which allowed Israeli transports to land at that air base once it was captured by IDF ground forces.

KABRIT (AB 228)

Alongside Abu Sueir and Fayid, Kabrit was the main EAF air base in the Suez Canal Zone. Traditionally that airfield contained an OTU which was No. 31 Squadron of 12 Air Brigade, flying MiG-17Fs. This unit was augmented by pilots that were on conversion to the Su-7 and hence it seems that its potential was high. Indeed, Kabrit was one of the three Egyptian air bases, the others were Abu Sueir and Bir Gafgafa, that were defended from the air as well as from the ground. That posed a problem to 105 and 109 Squadrons which were assigned Kabrit as their prime target. First to attack was a quartet of SMBDs from Hatzor that bombed the runway with two RPBs, each before the three strafing passes. During the strafing action, the leader, Seren Alexander Armon, observed a lone Il-14 coming in to land and shot it down. A second Il-14 was also shot down by that formation before it left the area. Then it was intercepted by a pair of MiG-21s, possibly from

Bir Gafgafa or Fayid. The number two man, Yak Nevo — the former C.O. of 101 and 105 Squadrons with three confirmed kills to his credit flying on an emergency posting — suggested the leader engage the Egyptian interceptors, but Armon ordered his formation to head east. Only two SMBD from that formation reached Hatzor, the leader, Seren Armon, and the number four, Segen Amiram Manor, were shot down and killed by the MiG-21s. Completing the work at Kabrit were another two formations from 109 Squadron. Both were engaged by MiG-17s but not a single Mystère was lost, while Seren (reserve) Asaf Ben-Nun, leader of the second formation, shot down an EAF MiG-17. Yet one more formation of SMBDs also claimed a single MiG-17 in air combat, raising to three the number of kills achieved by 105 Squadron over Kabrit, while the unit also claimed the destruction of a dozen MiGs on the ground, in exchange for two of its Super Mystères and pilots lost, both in air combat.

THE FIRST WAVE: A SHORT SUMMARY

About ninety minutes after the start of the attack, the first wave of Operation Moked was over. The goals of the operation to neutralize the EAF for the first critical hours of the war and to destroy a substantial part of its combat aircraft were achieved against a minimum cost to the attackers. The total claim was 204 EAF aircraft destroyed, including eight shot-down in air combat, and although that figure can be degraded a bit, to at least 140 aircraft, it was a magnificent achievement. The IDFAF lost during the first wave ten aircraft — five Israeli pilots were killed, two were taken POW and three rescued — representing a loss rate against the number of sorties flown of about 5% which was much lower than the anticipated figure. Of the ten losses, four were in air combat (two SMBD, one Vatour and one Mirage) meaning that the exchange rate was 1:2 in favor of the IDFAF which is quite bearable in an offensive action.

As a whole, the EAF was surprised. Most of its air bases were unable to scramble fighters to defend themselves and had to rely on AA fire that became more dangerous as the attack was pressed on since the initial shock was over and the Israeli aircraft paid no attention to defence suppression, concentrating instead in the higher value targets, combat aircraft. Only over three air bases, Abu Sueir, Bir Gafgafa, and Kabrit, the attacking formations were met by EAF fighters but their number was too small to change the outcome of the attack. The IDFAF plan was executed just as planned with one minor mistake, the attack of Cairo International which was not a designated target. The reports brought back by the pilots and the photos obtained by the reconnaissance Vatours and Mirages that took an active part in the attack, as well as those obtained by a Vatour IIBR sent specifically to photograph the Egyptian air bases, served the IDFAF planners at the Operations Department to plan the second wave for which targets were not assigned prematurely. Meanwhile the damage inflicted by the first wave meant that the expected EAF resistance was to be negligible. The stage was set for the second wave.

Mirage IIICJ number 725 in "clean configuration" takes off for a training hop. (IDF)

THE SECOND WAVE

After 09:00 it became apparent that the success of the first wave was overwhelming while casualties were very light when compared with the anticipated 25% loss rate. As a consequence, the art of damage assessment became crucial to ensure that no effort was to be wasted on targets that were already neutralized. At that stage there were still no indications that Syria or Jordan were to join the conflict and, hence, the logical conclusion was to press on the attack on Egypt — although part of the effort was dedicated to the attack of other targets than air bases, including radar stations, while quite a number of interceptors were kept on CAP to counter any Arab retaliation. Nevertheless, during the second wave, which lasted roughly between 09:00 and 12:00 hours, it became apparent that both Syria and Jordan joined the conflict and measures were taken to deal with the air forces of these nations. Thus, the second wave was different from the first mainly in the amount of planning. Many formations that were briefed to attack one target found themselves being diverted to an entirely different target, sometimes hundreds of miles away from the original one.

The navigation to the targets was easier during the second wave because the aircraft were flown at higher altitude, although generally the trend was to stick to the low altitude profile to avoid surprises, and the smoke columns from the previously attacked air bases were excellent landmarks. Naturally, the number of sorties flown in the second wave was smaller than that on the first, but the variation of less than 20% was not that significant, a total of 164 were recorded on the second wave. Just like in the description of the first wave, it is easier to understand the attack by a presentation according to the air bases under attack.

ABU SUEIR

Since the opposition at Abu Sueir during the first wave was fierce, it was plain that an effort to end the activity there would be launched. The Ouragans of 113 Squadron, the SMBDs from 105 Squadron and the Mirages from 119 Squadron were to visit Abu Sueir during that wave. This time, however, no aerial opposition was reported and none of the attacking aircraft was lost while at least one formation which was briefed to attack Abu Sueir, four Vatours of 110 Squadron, was diverted to another target even before the pilots manned their aircraft. In contrast to that episode, the 113 Squadron formation, which attacked Abu Sueir, was briefed to attack Fayid but was diverted in the air to Abu Sueir, which meant that the area was clear of EAF interceptors — otherwise the Ouragans would most probably not have been sent to the more distant target.

AL-MINAYAH

That air base was not attacked during the first wave and, hence, it is logical to suppose that it did not house any important elements of the EAF except for Il-14 transports but then it was possible that the base might be used to accept aircraft from other damaged airfields. Two Mirage formations were sent at high altitude, the distance involved preventing low altitude ingress, with the first from 119 Squadron led by Seren Oded Agee and the second from 101 Squadron led by Sgan Aloof Zoric Lev. No other aircraft than the Il-14s were observed and these were destroyed by the first formation during the strafing passes that followed the bombing. No targets were left for the second formation that nevertheless attacked Al-Minayah air base. A formation of Vatours from 110 Squadron was diverted on its way to Al-Minayah to Inchas.

BILBEIS

The home of the EAF Academy was also a probable sanctuary site and was attacked by a formation of Mirages from 101 Squadron that claimed the destruction of between six to ten Il-14 transports.

CAIRO WEST

The second wave attack on Cairo West caused the IDFAF its only casualty during that wave. A formation of 101 Squadron Mirages that attacked the air base was bounced by defending MiG-21s on its way back home and one of the Mirages was shot down. Its pilot, Seren Yair Nueman, ejected over Zagazig in the Nile Delta but was murdered on the ground by hostile civilians. Previous to the air combat, the formation claimed the destruction of at least two MiG-21s on the ground. Cairo West was also attacked during the second wave by a formation of 110 Squadron Vatours that took off at 10:23 after being given only five minutes to plan the attack. Once again the Vatours were engaged by SA-2s. The highly important Cairo West air base and its approaches were defended by five such batteries, but these did not cause damage while the formation had very little to add to the destruction of the airfield which was full of burning aircraft from the previous attacks.

FAYID

Just like nearby Kabrit, that airbase was certainly out of action by the first wave attack and as a consequence received less attention during the second wave. For example, a formation of 113 Squadron Ouragans on its way to Fayid was diverted in the air to Abu Sueir.

HELWAN

This air base did not contain operational units of the EAF but was a possible shelter for EAF aircraft from damaged airfields. Hewan was attacked by a formation of Mirages from 101 Squadron that claimed the destruction of two or three MiG-15/17s, a single MiG-19, a single Antonov transport and two lightplanes.

INCHAS

The first wave attack on Inchas virtually paralyzed the EAF top interception Air Brigade and, as a consequence, a marginal effort was devoted during the second wave to press a target with no offensive capability and almost non-existent defensive capability. A formation of SMBDs from 105 Squadron attacked the air base without any aerial opposition, although the SA-2 batteries that were stationed in the area to protect the nearby Egyptian nuclear plant were in action but totally ineffective. The SA-2 batteries were attacked by a formation of Vatours from 110 Squadron that was diverted to Inchas from Al-Minayah. The Vatours were unable to detect any worthwhile targets remaining and thus attacked an SA-2 battery at the initiative of the leader.

KABRIT

That air base was certainly put out of action by the first wave and thus it received less attention during the second wave. It was attacked by SMBDs of 105 Squadron and Mystères of 116 Squadron while a formation of four Mystères from 109 Squadron that were to take-off from Ramat-David at 10:15 hours to attack Kabrit was diverted to Amman, Jordan, as described later.

LUXOR

This air base in the Upper Egypt was the third in which EAF Tu-16s were based and while the other two, Cairo West and Beni Suef, were attacked during the first wave, it was impossible to attack Luxor and keep the secrecy of the operation because of the required profile of the flight. Thus a formation of three Vatours

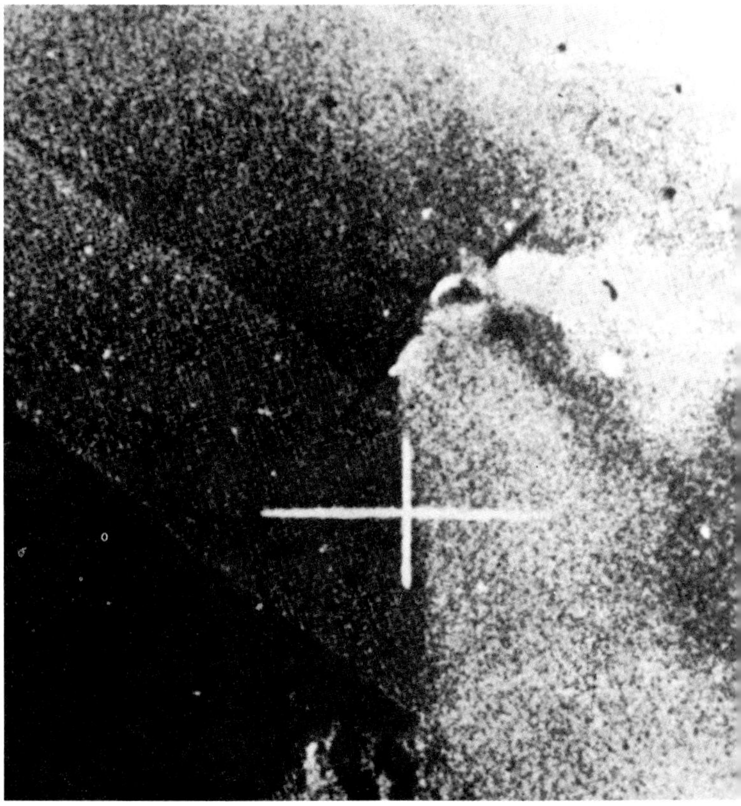

Streaming fuel from the pierced wing tanks, a Hunter plunges to the earth after having been vanquished by an Israeli Mirage. (IDF)

from 110 Squadron were dispatched to Luxor during the second wave flying an optimum cruise profile at high altitude which meant a steady climb as fuel was burned. As usual, the Vatours bombed the runways and then strafed the parked Tu-16s with no apparent opposition, not even AA fire. The formation claimed the destruction of eight Tu-16s and eight An-12s. Before leaving the target area, they photographed the air base for later damage assessment.

MENSURAH

This was an emergency dispersion site in the Nile Delta and, as such, was very difficult to locate. It was attacked during the second wave by a formation of SMBDs of 105 Squadron led by the unit's C.O. Rav Seren Shavit but it seems that the results were not positive as the same unit was to return to that target during the third wave.

By the early noon hour of June 5, as the second wave of Operation Moked ended, it was clear that the EAF was mostly paralyzed. With 107 aircraft claimed as destroyed being added by the second wave to the 204 of the first, and even if this figure is somewhat degraded to allow for overclaiming, the magnitude of the victory was almost unbelievable when compared with the light losses sustained by the IDFAF. Only nine aircraft were lost during the first and second waves of Operation Moked. Nevertheless, by that time Israel was absorbing counterattacks but not by the EAF, which was just not able to launch such attacks. The attackers were Syrians, Jordanians and possibly, just possibly, Iraqis.

The action against the IDFAF commenced just before noon with Jordanian artillery shelling Ramat-David, Tel-Nof, Lod International Airport and the former air base at Kfar Sirkin. At that stage it was not clear yet whether Jordan committed itself fully to the conflict or was just paying a debt to Egypt so a formation of Vatours, which was previously briefed to attack Abu Sueir, was ordered to attack King Hussein's palace at Amman as a signal for him to stay out of the conflict. Nevertheless, before that attack was launched, the magnitude of the Jordanian involvement became plain when RJAF Hunters attacked targets in Israel in parallel to a similar effort by SAF MiG-17s and MiG-21s.

The Syrian MiGs concentrated at targets in northern Israel while the RJAF Hunters attacked the center of the country. At least four targets, including the oil refineries near Haifa and the dam on the Jordan River near Deganyah, were attacked by the MiGs, but the important targets were missed altogether and the damage caused by these air raids was light. Only one of the SAF formations, the one attacking the dam near Deganyah, was intercepted by Israeli Mirages that claimed two of the attackers. Another air combat involving attacking SAF fighters was when a pair of MiG-17s tried to shot down a 100 Squadron Piper which they encountered in the Western Galilee. The Israeli pilot managed to evade the two Syrian fighters. One more SAF victim in that effort was a MiG-17 that was shot down by IDF AA fire at about 12:30 hours.

In parallel to the Syrian action, the RJAF Hunters performed a similar effort. They were slightly more successfully. The Hunters attacking Tel-Nof damaged a Dakota while those attacking Kfar Sirkin Camp destroyed a Noratlas transport. Interestingly enough, Iraqi newspapers of June 6 claimed that IrAF (Iraqi Air Force) Hunters were the ones that attacked Kfar Sirkin and, in the absence of any other evidence, it seems that Iraqi fighters were already forward-based in H-3 airfield. At least four other targets were attacked by the Jordanian or Iraqi fighters. Here too, only one of the attacking formations was intercepted by a pair of 119 Squadron Mirages, with the leader, Seren Oded Sagee claiming one of the RJAF Hunters. So the third wave, that began at around 12:00, involved attacks on Egyptian, Jordanian and Syrian air bases while even a single Iraqi air base was attacked toward the end of that wave. The attacks on Eben Al-Walid (H-3) was executed since it was though that Iraqi Hunters attacked Israel. This happened shortly after the RJAF air bases were attacked and thus the claim that the attackers were Iraqis.

Moreover the targets attacked were in the traditional Iraqi slice of the front, between that of the Syrians and the Jordanians. The Hunters attacked Ramat-David air base, which they missed, dropping their bombs on a nearby settlement, and the IDF ground forces concentrations facing the Shomrom, where considerable damage was caused. For those two reasons, it is claimed that the Hunters were Iraqis and not Jordanians, but until this day it has not been proved.

THE THIRD WAVE

Since the third wave involved an attack on three air forces that were still intact and well aware of what was going on, the surprise could only be achieved tactically, involving the direction of the approach, the time of the attack, the number of attacking aircraft, etc. Even this could not ensure a casualty-free attack and, indeed, the third wave saw a climb in the number of losses when compared with the second wave, which was basically an attack on a mostly destroyed and defenseless air arm. The description of the third wave follows the usual pattern, but includes (after the Egyptian), Jordanian, Syrian and Iraqi air bases.

ARDAKA (286 AB)

Housing a MiG-19 unit plus helicopters, that air base was seen as a threat to any Israeli attempt to capture Sharm El-Sheikh and open the Straits of Tiran to Israeli shipping and air traffic. Thus, two formations of Mirages from 119 Squadron were dispatched to attack this air base flying in a straight line at 30,000 feet. The first formation, led by the unit's C.O. Rav Seren Ran Peker, descended to low level above the Red Sea, since it was clear that the defenses at Ardaka were fully alert, being welcomed by intense flak after climbing for the identification and bombing. Nevertheless, the formation pressed on the attack until it was intercepted by a quartet of MiG-19s which were possibly on CAP waiting for the Israeli intruders at the wrong place and hence taking them quite a while to return to their air base.

Ignoring their low fuel state, the Mirages met the MiG-19s, with the leader shooting down two and his wingman another one. As the Mirages left Ardaka, they witnessed the remaining MiG trying to land on the cratered runway but exploding in the process. It was clear that Ardaka was closed and shocked, so the second formation was simply diverted to Amman air base when it was near Sharm El Sheikh.

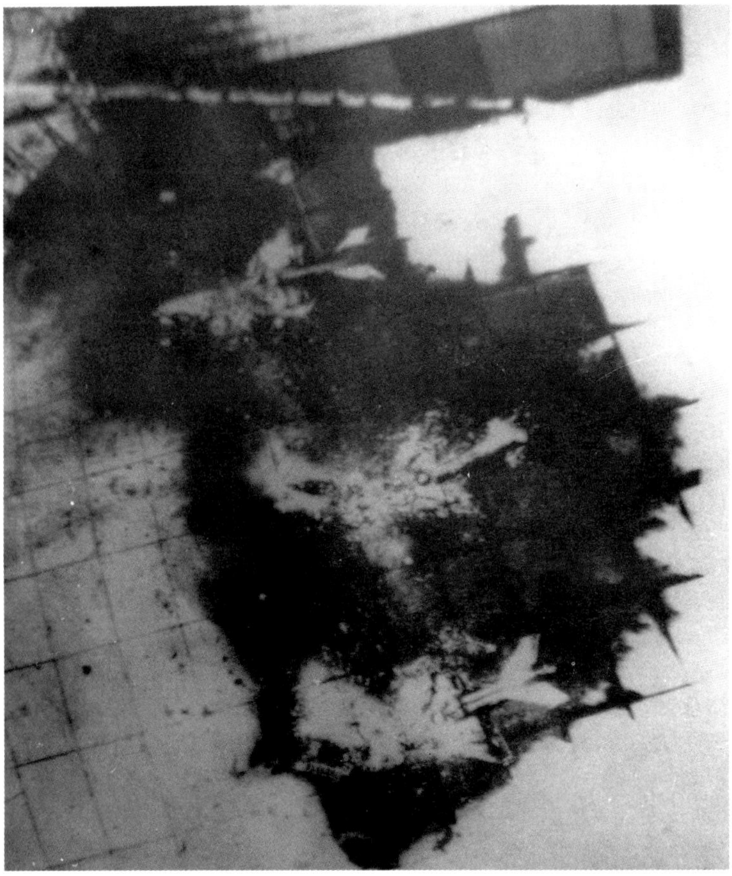

Dramatic image of three Egyptian MiG-21 interceptors destroyed at an Egyptian air base during the first wave of attacks on June 5, 1967. (IDF)

BIR TMADA

A pair of Vatours attacked that air base as well as Egyptian armor at Jabel Libni on their way back to base. They also photographed in the desert an Il-14 after a crash landing, most likely the one that was claimed by the Mirages of 101 Squadron near Jabel Libni earlier that day.

INCHAS

Although basically out of action, it was important to keep the pressure on such a high priority target and to verify that it was indeed out of action. Thus the target was again listed for attack by 105 Squadron, although this time the SMBDs were never to reach Inchas as they were diverted to another target at Jordan when they were more than halfway to it.

MANSURAH

First visited during the second wave by a Super Mystère formation from 105 Squadron which continued the attack with two more formations, one of which failed to locate the air base and performing a CAS (Close-Air-Support) at Sinai instead.

RAS BANAS

Even further south than Ardaka, and also on the shore of the Red Sea, Ras Banas could have been a shelter for the remaining EAF aircraft. It was attacked by the only type in the IDFAF arsenal that had the sufficient range performance, the Vatours of 110 Squadron. Thus, during the third wave attack on the EAF, none of the attacking IDFAF aircraft was lost while attention was given to two remarkably distant air bases to complete — as much as possible — the destruction of the EAF; although the effort against this air arm during that wave was on a lower scale as at about 12:45 hours the first formations to attack the RJAF air bases were launched.

AMMAN

The air base was part of Amman International Airport. It was the home of the RJAF support elements, the transports and helicopters. As such, its importance in comparison with Mafraq was negligible but nevertheless it was attacked mainly by formations that were already in the air on their way to other target. The first to attack were four 119 Squadron Mirages which had been diverted from Ardaka. Another formation to attack Amman as a result of an inflight diversion was composed of four 116 Squadron Mystères led by Seren Avihu Ben-Nun. The number three of that formation, Segen Dror Harish (a future 11-kill ace), was injured in his left shoulder by a round that entered his cockpit and went out through the canopy. He landed at Tel-Nof. In addition to the military aircraft, the Mystères also destroyed an Alia DC-7.

Another quartet of Mystères, albeit from 109 Squadron, was briefed to attack Amman instead of Kabrit but eventually attacked neither as it was diverted to Elmaza, Syria, while a formation of 107 Squadron Ouragans did attack Amman. A 101 Squadron formation also attacked Amman. No IDFAF were lost over that air base.

MAFRAQ

The home of the two RJAF Hunter squadrons was a different story and here the IDFAF lost several aircraft. As usual, by that time the formations that attacked Mafraq were mostly diverted to their new target rather than given the appropriate time and tools to learn the target and plan the attack, it was a kind of catch: get a target, go and take it out of action. Either because it was close or thought to be an easy target, the Ouragans and Mystères were sent to knock out Mafraq, a job they did very well but at a cost of two pilots, one flying an Ouragan and the second a Mystère.

Seren Yoram Harpaz, flying an Ouragan, was thought to have been hit by AA fire, but was blasted out of the sky by an IDFAF Hawk SAM from the battery that defended the nuclear plant at Dimona. Harpaz did not respond on the radio and since the Dimona area was strictly a prohibited zone, no chances were taken and the Hawk was launched. On the other hand, Segen Hananyah Boleh was shoot down in his Mystère by a RJAF Hunter over Mafraq.

Practically, by 14:45 the RJAF ceased to exist. Its serviceable Hunters were destroyed, as most of its transports and helicopters, and its two air bases were severely damaged. In parallel to the attacks on Amman and Mafraq, a similar effort was launched to knock out the SAF with considerable success although, naturally, the element of surprise was lost and the SAF suffered heavy losses but was not wiped out.

DUMIER

This air base east of Damascus was surprised by a formation of four SMBDs from 105 Squadron that bombed the runways and strafed the MiGs; but, the defenses overcame the initial shock and two aircraft were lost, one each of the following formations.

Segen Dan Sigri of 105 Squadron was murdered on the ground by civilians while Segen Avraham Weiland of 110 Squadron, flying wingman to the unit's C.O., was taken POW. Another 110 Squadron formation followed the one which lost Segen Weiland while two formations of 109 Squadron also attacked Dumier with the first, diverted from Marj Real air base, claiming the destruction of eight aircraft and the second adding two MiG-17s and several helicopters to the total claim.

ELMAZA

Located on the outskirts of Damascus, Elmaza was the birthplace of the SAF. The first formation to attack this air base, which housed MiG-17s, was composed of four 109 Squadron Mystères that were originally intended to attack Kabrit, then instructed to attack Amman and finally ordered to attack Elmaza. In order to achieve a tactical surprise, the four aircraft flew north along the Lebanon Valley, crossed the Hermon Mountains ridges over the Beirut-Damascus road and then to Damascus. On their way, the number two had to abort due to a fuel transfer problem and the remaining three bombed the runways and strafed the MiG-17s. No aerial opposition was encountered but the flak was intense and Segen Nachom Merchavi, number four in the formation, received a hit but managed to eject over Lebanon where he was taken POW. Other formations attacked Elmaza as well, including one that was diverted from an interdiction at Sinai, but no more losses were recorded over this air base during the third wave, although more aircraft were lost during the fourth wave.

TSAIKAL

The first formation to attack that air base did its job but reported to the second formation that SAF MiG-21s were defending the air base. That second formation was a quartet of 105 Squadron SMBDs led by the unit's C.O., Rav Seren Aharon Shavit and it had problems, no doubt. Since the formation flew in a straight line to Tsaikal from Hatzor at low level, they overflew the Jordanian held Shormrom area where small calibre fire caused a minor but continuous fuel leak to the number three SMBD. Nevertheless he pressed on and they all managed to bomb the runways but immediately after were bounced by a pair of MiG-21s. Number three, with no fuel for an air combat, climbed for height while the leader engaged one MiG and numbers two and four the other Fishbed, eventually shooting down both, with a funny incident involving the second kill. After the leader shot down his MiG he heard his wingman on the radio, a young pilot just incorporated in the squadron, saying "I have the MiG right in front of me, what I shall do now?", the squadron commander yelled "press the trigger, you idiot!" and was replied "Wow, he sheared off its wing." Not forgetting their purpose, the SMBDs did their strafing passes, destroying three Ilyushins and four MiG-17s before leaving the area.

MARJ REAL

This air base was attacked by 105 Squadron SMBDs followed by 107 Squadron Ouragans. A quartet of 109 Squadron Mystères that took off to attack Marj Real at 13:45 was diverted to Dumier as reports claimed that no viable targets remained at Marj Real. No losses were suffered over this airfield.

The IDFAF lost three aircraft over Syria during that wave, a Mystère, a Vatour and an SMBD, but the Syrian losses were considerably heavier. Finally, almost as a finishing touch to the third wave, the Iraqi H3 airfield was attacked by a formation of three Vatours led by Seren Gideon Maggen and Seren Alexander Inbar in a Vatour IIN since the number three had aborted. When they climbed to identify the target, they were intercepted by two IrAF MiG-21s but, nevertheless, managed to bomb the runway. The attack was by then not as planned, hence, the leader dropped all his bombs on his first bombing runs, while the plan called for two bombing runs, and on the near (when observed from the west) third of the runway instead of the furthest because he saw two Hunters about to take-off. Then he made two strafing passes, the first on a mixed line of MiGs and Hunters and the second on a transport aircraft, while his number two first strafed the two Hunters about to take-off and only then bombed. Number three bombed the airfield with only half the load of his bombs, sticking to the original plan, and so had to jettison the remaining bombs over the desert on their way back. Amazingly, throughout the attack the IrAF MiGs were unable to shoot down any of the Vatours while the leader, Gideon Maggen, stated that if they were late by several minutes and the two Hunters were in the air, they would probably all have stayed in Iraq, expressing the respect that the Israelis felt for the British fighters.

So by the afternoon of June 5, the IDFAF was the unrivaled ruler of the Middle East skies. The EAF and SAF were paralyzed, although not totally destroyed and their remaining aircraft would be encountered in the coming days, the RJAF was practically wiped out, the IrAF was driven into a defensive posture and the Lebanese Air Force received a signal to stay out of the conflict. That incident happened during the early noon hours after Iraqi or Jordanian Hunters attacked Ramat David. Two pairs of 117 Squadron Mirages were scrambled but failed to intercept these Hunters and instead flew CAP. One of the pairs was vectored by the controller north to the Lebanese border where a lone aircraft was observed to intrude the Israeli air space. However when the Mirages got nearer, it retreated but intruded again when the Mirages left the area. That happened several times until the Mirages finally received permission to cross the Lebanese border. The leader, 117 Squadron vice commander Seren Ori Aven Mir, shot down the LAF Hunter over Rayak air base, getting amazed in the process by the Hunter's ability to absorb punishment and still continue to fly. Other activity during the second wave was an attack on a Jordanian radar station by 105 Squadron, while the tale of the IDFAF attack effort on June 5 will be described separately after the description of the fourth wave completes the story of Operation Moked.

THE FOURTH WAVE

Beginning at around 16:00 and continuing until darkness, the fourth wave concentrated on the SAF air bases while no formations were launched to attack IrAF or RJAF air bases and a small effort was launched against the EAF. It must be stressed that by the afternoon quite a percentage of the IDFAF sorties were directed to support the ground forces as air superiority was indeed achieved. As in the third wave, no losses were sustained over Egypt during the fourth wave. Cairo International was attacked again, this time by 119 Squadron Mirages led by Rav Seren Peker, while another formation from that unit was diverted to T-4 air base in Syria. The nearby Cairo West was taken care of simultaneously by a 101 Squadron formation in which Sgan Aloof Dani Shapira flew as number three. Several SA-2s were launched at them but they completed their missions uneventfully, attacking a radar at Romani on their way back home. SA-2s were also fired against a 109 Squadron formation on its way to Kabrit and that same formation identified another SA-2 battery northeast of the Great Bitter Lake and attacked its radar and generators. In sharp contrast, the attack on SAF air bases cost the IDFAF three fighters, exactly the number lost on the third wave. Two 119 Squadron formations attacked T-4 with considerable success and the first formation even claimed the two defending MiG-21s with the kills credited to Seren Asher Snir and Segen Giora Rom. Flying in the second formation was Segen Menahem Shmul, whose Atar flamed out. Luckily, he managed to re-start it. The Israeli losses over Syria during that wave were all attributed to the massive AA fire in the vicinity of the capital, Damascus, where two air bases were located, Elmaza and Damascus International Airport. These were attacked by formations of 101, 105, 116 and 117 Squadrons with the last being particularly unfortunate. The leader of the quartet and squadron C.O., Rav Seren Amihai Shmueli, had to eject over Israeli territory. He was injured and temporarily replaced as 177 Squadron C.O. by Seren Aven Nir. His wingman Segen Meir Sahar was even less fortunate. He received a direct hit and was killed just a few hours before his brother, Rav-Seren Yonatan Shahar, was rescued from Egypt. The third loss involved a 116 Squadron Mystère with the pilot, Segen Yonathan Zoreah, losing his life.

Thus ended Operation Moked. Its objectives were achieved at the cost of twenty Israeli aircraft lost, of which ten pilots were killed, two were murdered after ejecting from their crippled aircraft, four were taken POW and three were rescued. Four of the aircraft lost were Mirages, two from 117 Squadron and one each from 101 and 119 Squadrons. Thus the IDFAF lost about 10% of its first line combat aircraft, still below the anticipated 25%. Losses were shared by all the front line units, most of them were to AA fire, but five were in air combat, and a sixth to a related cause of fuel starvation; but the claimed kills, 23 Arab aircraft, overweighed the losses, especially when it is considered that Operation Moked was an offensive operation. Clearly leading the score was 119 Squadron with ten kills, while 101 Squadron scored three and 117 only one; the remaining were claimed by Super Mystères, Mystères and Ouragans. However, although these were indeed all the kills recorded by the IDFAF during the first day of the war, it was not so as for losses. In parallel to Operation Moked, the IDFAF supported the ground forces with the percentage of sorties devoted to such missions steadily rising as the success of Operation Moked was revealed and fewer sorties were required to complete the damage.

THE IDFAF ON JUNE 5

Since all the combat elements of the IDFAF were committed to Operation Moked, it was clear that the IDFAF ground forces would have to do without air support at least during most of the first day of fighting. Nevertheless, careful thought was given to the subject of CAS and it was decided to use the 12 Flying School Magisters in that role. In addition, all other support elements of the IDFAF were closely involved with the ground fighting right from the start. The first action of 12 FS Magisters was the deception early in the morning already mentioned early. Then the Magisters were armed and dispatched to attack ground targets at Sinai, especially radar stations and artillery batteries but also command posts, in parallel to the beginning of Operation Moked. Because of the Magister's relatively low speed and the absence of an ejection seat, it was obvious that using the jet trainers to attack well defended targets was a risk, but it was concluded that such a risk had to be taken as this was the only possible way to provide CAS during the first crucial hours of the conflict.

The first formation of Magisters were sent to attack radar stations, artillery batteries and El-Arish air base at around 07:45 to 08:00. During that action the unit lost its vice-commander, Seren Arnon Livnat, but continued throughout the day with CAS and BAI (Battlefield Air Interdiction) missions. Until noon the war was on the southern front where three IDF divisions were stationed along the Egyptian border. While providing CAS to the southernmost division facing Abu Ageila, a Magister was lost to Egyptian AA fire with its pilot, Seren Rafael Elisha, killed. As a consequence, the division C.O. Aloof Sharon was ordered to cease all CAS action in his area of command so the morale of his soldiers would not be affected if another Israeli aircraft would be lost in front of their eyes. At that stage, the division was still statically shelling the Egyptian forces in anticipation to the night action. Another Magister was also lost in the south and again the pilot, Segen Matityahu Buremboim, was killed. Another Magister was almost lost when Seren (reserve) Ya'acov Morgan, formerly an IAI test pilot and then an El Al captain, brought back to base his damaged Magister. He was leading a formation to attack the Rafah-El-Arish road where an artillery convoy was observed. The Magisters attacked the towed artillery pieces. On the first pass Seren Morgan Magister was hit in the left aileron. Using the trim tab, Morgan made a second attack run before returning to Hatzerim where his second landing attempt was successful. The Magisters were supported in their activity over northern Sinai by 107 Squadron that, after contributing three formations to Operation Moked first wave and two more formations during the third wave, concentrated on attack missions at northern Sinai. The squadron attacked artillery batteries and radar stations.

During the noon hour, more and more first line combat aircraft were directed to attack ground targets at Sinai and hence the Magisters were given Jordanian targets in the central front. Nevertheless, not all the aircraft ordered to attack in Sinai did so as quite a number of formations were diverted to attack RJAF and SAF air bases. The targets that were attacked during noon included the Jordanian 40th Armoured Brigade that was on the move from Adam Bridge south to Yerihu. In addition radar stations and AA sites in Jordan were also attacked, while by the early afternoon hours the focus shifted to the Yerihu-Jerusalem road where many BAI missions were flown from 17:30 hours until 18:45. A single Magister was lost in that action when Seren (reserve) Shabtai Ben-Aharon was forced to abandon his aircraft. An S-58 captained by Seren Yehuda Perri was sent to retrieve him, with a pair of 101 Squadron Mirages returning from Amman acting as cover, but was only able to locate the parachute before leaving the area damaged by seven direct hits. It was later learned that the Israeli pilot was captured by Jordanian soldiers who burned him to death. Slightly earlier, the Iraqi 8th Brigade, which was the spearhead of an expeditionary force to Jordan, was attacked by the formation of three Vatours that returned from H-3. Seren (reserve) Ran Goren, the number two pilot in the formation, strafed the Iraqis.

An interesting attack mission carried out at twilight involved Seren (reserve) Gideon Maggen during his third sortie of the day. Maggen led a formation of Vatours to attack the elite Shazli Force at Wadi Couriyah in Sinai but the navigation and target identification was problematical due to the falling darkness. Finally Maggen, who flew alone in Vatour IIN number 66, observed an Egyptian convoy and came in to attack it. The Egyptians reacted very quickly. They all switched off their lights but this did not conceal them as they all stopped. The brake lights merely replaced the headlights as an identification landmark to the IDFAF pilots. But the convoy also opened fire and Maggen received several hits in the belly of the Vatour, damaging the fuel tanks, and the nose, where the hydraulic system was located and was also damaged. He landed at Ramat-David but, as the damaged Vatour slowed down, the aerodynamic forces diminished and, having no hydraulic power to steer his aircraft, Maggen simply let it roll off the runway into a field where it stopped. Next day he was flying again in the same Vatour IIN (66) which was fixed by the ground crew that night, only for the AA fire to damage it once again. Once more the aircrew accepted the challenge and by June 7, the aircraft was again in the air.

By the late afternoon and early evening of June 5, the IDFAF was able to devote its first line aircraft to the attack role since Operation Moked was practically over. The primary target at that stage was the Jordanian artillery which shelled many targets in Israel including Tel-Aviv. 101 Squadron Mirages attacked an artillery battery near Janin just before darkness. Later the fighters, most usually SMBDs or Vatours which had the necessary combination of good warload and effective endurance, orbited above the suspected area and attacked the batteries immediately after they fired. Obviously such a tactic did not result in the complete destruction of the Jordanian artillery, but it somewhat degraded the threat by placing constant psychological pressure on the Jordanian soldiers.

The Syrians also opened artillery fire that evening but were more clever than the Jordanians and stopped shelling once they heard that the 109 Squadron Mystères had launched at 19:39 to silence them. Thus the jets were diverted to the Jordanian front, attacking an artillery battery near Kalkilyah. While the combat aircraft and the reconfigured trainers were actually delivering their load to enemy targets, the other elements of the IDFAF were extremely busy doing their valuable jobs. Especially hard working were the three helicopter units flying transport, assault, CASEVAC, observation, communications relay and SAR. Thus the first day of the Six Day War was not only the show of the frontline fighters, it was a much broader effort. In addition to the helicopter, the transports and lightplanes also took an active part. June 5 was over and the achievements of the IDFAF were almost unbelievable.

However successful Operation Moked was, it must be remembered that it was only the opening act of a full scale war. At Ramat-David the 1st Wing C.O. Aloof Mishne Yehezkel Somekh gathered together all the wing's aircrews that evening and told them that Israel had already won the war, but from now on they should be more careful. He was certainly right and Seren (reserve) Gideon Meggen was convinced that Somekh's far sight probably saved the lives of at least several pilots.

THE IDFAF ON JUNE 6

As the nature of equipment used at the time dictated that night operations were strictly limited, the IDFAF mainly operated during the day. Nevertheless, an effort was made to fully explore the air superiority even at night and elements of the Jordanian 27th Infantry Brigade and 60th Armoured Brigade were attacked during their nocturnal move from Yerihu to Jerusalem. Illustrating the problematic nature of night operations at the time was the attack by three Vatours of 110 Squadron during which one of them released flares while the other two bombed. Another 110 Squadron formation that took off at 02:26 to support the Sharon Division in its nocturnal combat at Om Kattef was diverted to the Yerihu-Jerusalem road after failing to identify between friends and foes at Sinai. They too bombed the area and returned to Ramat David at 03:45. Obviously such a nocturnal attack tactic did not ensure

Test pilot Danni Shapira was instrumental in the development of the Mirage 5J fighter bomber upon the IDFAF specifications. But the 50 aircraft were embargoed by the French government in 1968 ending in Armée de l'Air service as Mirage 5Fs. Meanwhile Israel developed its own Mirage 5J —with unofficial French help—, the IAI Nesher, and from it derived the more capable Kfir. (Dassault)

the destruction of the Jordanian forces but it undoubtedly exerted a lot of psychological pressure on the Jordanian troops. That interdiction effort was intensified during the daytime while CAS sorties were also flown on both the Egyptian and Jordanian fronts.

On the Jordanian fronts, the IDFAF operated throughout that day in the Shomrom area, where the IDF Armoured forces were engaged with the Jordanian 25th Infantry Brigade and 40th Armoured Brigade, and in the vicinity of Jerusalem, where the interdiction efforts of the IDFAF bore its fruits when Eastern Jerusalem was captured on the next day, June 7.

The air force involvement commenced at dawn when 109 Squadron sent its Mystères for a CAS mission at Kabatiyah Junction in the Shomrom. The jets located the Jordanian tanks, hidden in the olive groves, and claimed the destruction of twelve, which is obviously too high since the total Jordanian losses at that junction until noon were fourteen M-48 Patton tanks. Two quartets of 107 Squadron Ouragans also operated on CAS missions at that area while further south, east and south of Jerusalem actually, the Magisters of 12 Flying School continued with the interdiction effort which they started the previous day. The Mystères of 109 Squadron and the Vatours of 110 Squadron also operated in that mission as well as flying CAS missions in the vicinity of Jerusalem together with Ouragans. The Magisters also flew CAS missions in Jerusalem that morning and Segen Dan Giveon of the formation that took off at 10:51 to attack a battery of mortars at Mar Alias was shot down and killed by AA fire. That was Giveon's third sortie of the day, the previous two were interdiction missions east of Jerusalem, take-off times were 06:10 and 08:20 which gives an indication of the rapid turn around periods achieved by the ground crews.

Just 21 minutes prior to the take-off of the above mentioned formation, the C.O. of the Jordanian 40th Armoured Brigade asked for urgent air support but was answered that this was not possible. The Jordanian High Command did not bother to inform the C.O. that the RJAF existed no more as a fighting element, although that officer was well aware of the IDFAF activity. The Brigade suffered heavily from Israeli air strikes while being pressed by superior IDF forces attacking from several directions. The fruitless movement of the unit from the Jiftlik area to Yerihu the previous night also took its toll, and the brigade was very low on fuel and ammunition but the seven fuel and ammo trucks that were sent to resupply the unit were destroyed by an Israeli air raid. Hence the Jordanian units in the Shomrom were ordered to retreat east at 15:30, thus effectively drawing to an end the fighting in that area. Down south fighting continued in the Jerusalem area but here too the Jordanian forces were doomed. At 18:00 IDF forces captured Calandyah airfield north of Jerusalem and by 22:30 the IDF Central Command asked for an interdiction effort east of Jerusalem in preparation for the final battle next day.

Throughout Day 2, the IDFAF mainly flew interdiction and CAS missions on the Jordanian front, but helicopters were also used for transportation and CASEVAC. Only a single aircraft, Segen Giveon's Magister, was lost on that front during June 6, although the flak was indeed heavy and it is claimed that every single 107 Squadron Ouragan sent for a mission over the West Bank returned with some degree of battle damage.

On the Egyptian front, the IDFAF also concentrated on interdiction and CAS missions in Sinai. IDFAF pilots also flew CAPs, as the EAF was not totally wiped out, which indeed helped to waive several EAF air raids. The IDFAF helicopters were especially active in that front on CASEVAC, observation, transport and SAR missions. During the morning hours EAF Su-7s tried to attack the IDF Tal Division, operating in northern Sinai, but were intercepted by Mirages of both 101 and 119 Squadrons. Seren Oded Sagee of the latter was credited with his second kill while Segen Giora Epstein of the former unit was credited with the first of his eventual total of seventeen, earning the title of world's top jet ace. Three more Su-7s were claimed by other 101 Squadron pilots but one of the Mirages was lost in combat. In the same area that morning a pair of 119 Squadron Mirages on CAP intercepted four EAF MiG-21s that tried to surprise four Mirages returning from an attack mission. Segen Yeari shot down one of the MiGs with cannon fire after his wingman Shafrir AAMs missed. In yet one more air combat, albeit during the afternoon and above Jebel Marrah, Seren Sagee of 119 Squadron was credited with his third kill when he and his wingman shot down two EAF MiG-19s. Another two MiG-19s were claimed at the same area by pilots of 101 Squadron, one of them flying Mirage number 159 that registered its third kill. The first kill was an SAF MiG-21 on July 14, 1966 and the second an EAF Il-14 the previous day, later to reach a total of twelve kills plus an AS-5 missile.

While the IDFAF interceptors were denying attack opportunities from the remaining elements of the EAF, the rest of the IDFAF combat aircraft were exercising complete air superiority over Sinai. The Egyptian Army 3rd Division was attacked since dawn to frustrate its effort to advance toward the IDF forces in the El-Arish area in a BAI action. CAS missions in Sinai were performed by SMBD, Vatours that concentrated at the Gaza Strip, Mystères and Ouragans. A single Ouragan of 107 Squadron was lost when supporting the Sharon Division at Kseimeh. Its pilot, Rafael Lev, ejected over hostile territory and was rescued that evening by a 124 Squadron S-58 aided by a communications relay C-47 after a similar effort during the hot noon hours failed. Also at Kseimeh a 105 Squadron SMBD was lost but its pilot, Segen Yair Barak, was less fortunate and he was taken POW. A notable event was the first landing of a Noratlas transport at the captured El-Arish air base just before darkness.

The IDFAF action in the Egyptian front on Day 2 was highly successful. It fully explored the benefits of the air superiority that was gained during Operation Moked with combat aircraft flying interdiction, CAS and CAP missions and the helicopters supporting the advancing forces. Losses in that front on Day 2 were light, with three aircraft, while in air combat eleven EAF aircraft were claimed by 101 and 119 Squadrons.

While on the Jordanian and Egyptian fronts the IDF was on the offensive, the opposite was true in regard to the Syrian front where the IDF forces were in a defensive posture. The Syrians mainly shelled the Israeli strong points, concentrations and civilian targets, fully exploring the topographic advantage offered by the Golan Heights — which is some 800 meters above the Israeli held Hula Valley. The full support of the IDFAF was required here, both as a flying artillery and to frustrate the only Syrian offensive

Air-to-air view of Mirage IIICJ number 159, later in its operational life, while serving with 190 Squadron before being delivered to the Argentine Air Force, in 1995 it continued flying as C-711. (IDF)

effort that day when an armoured column attacked kibbutz Dan but retreated after being attacked from the air within hundreds of meters from its target. That morning, a formation of four Mirages bombed the important Syrian command post at Kuneitra with one of the Mirages damaged by the intensive flak. During the air attack in the Golan Heights, the IDFAF pilots faced a remarkable concentration of AA guns and many aircraft were damaged although only one was lost, that was a Vatour of 110 Squadron with its pilot ejecting over Galilee.

The prime targets attacked in the Golan Heights were strong points, command posts and artillery batteries, and that action mainly involved the Vatours and Mirages. The latter attacked a command post at Kuneitra and a strong point above Bnot Y'acov Bridge on the Jordan River, during which the leader, Segen Giora Rom of 119 Squadron, received a direct hit from a 37mm AA gun. Rom was injured and his Mirage severely damaged, but he managed to bring it for an emergency landing at Ramat David AB. Also considering an ejection was Seren Gideon Maggen who was leading a formation of Vatours to support an Israeli Navy commando raid at Latakiyah port. That mission was canceled, however, and the formation was directed to attack in the Golan Heights. A single AA shell struck Vatour IIN just between the cockpits, knocking off Maggens' ability to communicate with his navigator, Seren Menahen Ainee. Since the run was from east to west, Maggen hesitated in the hope that Israeli territory would be reached before ejection. However, when he turned off the pressurization, the smoke disappeared and Maggen was able to bring number 66 back home to Ramat David.

That was one of the four sorties that Maggen flew that day, three were attack missions at the Golan Heights and in the fourth he attacked a Jordanian armoured column. Hence, the IDFAF managed to keep the Syrian Army static at its positions at the cost of a single Vatour whose pilot was retrieved while another Vatour was damaged but its pilot, Seren Rafaeli, reverted from the primary hydraulic system to the mechanical back-up and made a successful belly landing at Ramat David.

Further east than the Golan Heights, another front was opened the previous day with Iraq when 110 Squadron Vatours raided H-3 air base. That air raid was not conclusive and it was decided to dispatch another attack on the morning of Day 2. Nevertheless, the Iraqis preceded the Israelis this time and early in the morning dispatched a quartet of Tu-16 bombers from Habbaniyah air base. It is claimed that two of the bombers returned to the base shortly afterwards, the third landed enroute at H-3, and only the leader, with the unit C.O. aboard, pressed on its attack. The Iraqi bomber dropped its bombs on the city of Natanyah, north of Tel-Aviv, and probably tried to return home at low level through the Izra'el Valley but that route placed him exactly over Ramat David. The pilots of the Vatours were at their cockpits when they saw the Iraqi bomber overflying their base with a pair of 117 Squadron Mirages on its tail. The Israeli pilots, Segen Ahud Hankin and Segen Amnon Arad, each launched a Shafrir AAM at the Iraqi bomber. These only damaged the Tu16, which was brought down by an L70 battery that defended the air base. Unfortunately the Tupolev crashed at an IDF barracks near Maggido, killing more than ten soldiers in the process. Shortly after the Tu-16 story ended, the four Vatours took off with their uncoordinated escort, a pair of 117 Squadron Mirages led by Seren Yehuda Coren. Either the Iraqis were warned of the attackers arrival or by coincidence the IDFAF aircraft arrived exactly when the Iraqis were replacing the Hunter CAP with a pair of fresh MiG-21s that were taking-off just as the Israelis arrived. The Vatours bombed and strafed the air base, while the Mirages battled the Iraqi fighters that tried to shoot down the Vatours. As a consequence, the attack was not that successful; but, in the air combat the Israeli success was overwhelming, with Seren Coren shooting down a Hunter and a MiG-21 and the number three Vatour, Seren Ben-Zion Zohar in aircraft number 17, shot down a Hunter. Finally Coren ordered its retreat back home. On their way back the Vatours exhausted their ammunition on the Iraqi 8th Brigade which was sent as an expeditionary force to the Jordanian front. Thus Day 2 ended, with fourteen Arab aircraft shot down, while Israeli losses amounted to five aircraft and one pilot.

THE IDFAF ON JUNE 7

The situation on the beginning of Day 3 was that IDF forces were penetrating even deeper into Sinai but only on its northern part. The remains of the Jordanian Army were retreating from the Shomrom area and the Syrian Army was still statically shelling Israeli targets. The two major operations planned for Day 3 were the capture of Eastern Jerusalem and its surroundings, thus completing the conquest of the West Bank, and an airborne assault at Sharm El-Seikh to capture the strategic point at the southern tip of the Sinai. In addition, the IDFAF supported the advancing forces in Sinai and bombed the Syrian positions at the Golan Heights. Hence it is impossible to point out a single concentrated effort on Day 3 as the IDFAF was stretched to its limits and was well involved on all fronts, plus an unfortunate attack at the fourth front of Iraq.

Up north the Syrian front was static but they continued to use their topographic advantage and numerical superiority in artillery pieces. The IDF could not challenge the Syrian artillery and hence the IDFAF continued its effort as a flying artillery, attacking not only batteries, but also strong points at the border and command posts deeper inside the Golan Heights in preparation for a possible IDF offensive operation. This offensive operation was not decided upon yet as the army was fighting on two fronts and to open a third was undesirable; but, the success on those two fronts gave support to the belief that once offensive operations end on one of the two active fronts the center of the operation would shift north. The IDFAF attacks at the Golan Heights were flown repeatedly and unglamorously against a well defended force, and once again the topography of that mountainous area allowed a concentration of AA guns at a strictly limited extension, while the attacking aircraft had very few options from which to arrive and attack. Moreover, the presence of low clouds over the target presented the Israeli pilots with even a greater challenge as identification was lengthy and dangerous. Only two air force aircraft were lost that day at the Golan. The first was an SMBD of 105 Squadron attacking artillery posts on the central part of that front, while the second was a Mystère of 116 Squadron whose pilot, the IAI test pilot Seren (reserve) Asaf Ben-Nun, ejected exactly over the frontline, but aided by an eastern wind landed barely within the Israeli lines. The IDFAF effort continued well into the night and, in spite of the primitive nocturnal attack equipment and tactics, a pair of 109 Squadron Mystères, led by Seren (reserve) Aryeh Dagan, were able to silence an artillery battery that shelled Rosh Pina. If not destroyed, the battery crew were at least frightened enough to cease their fire.

A bit to the south heavy fighting continued at the West Bank throughout Day 3. During the night the IDFAF continued to attack Jordanian forces and, most significantly, destroyed the command vehicle of the Jordanian Army 40th Brigade C.O. who lost all radio communications with his forces and retreated by foot to the eastern bank of River Jordan. When daylight broke, the IDFAF effort on that front was divided between the support of the IDF forces completing the conquest of the Shomrom and the support of the fighting forces at Jerusalem.

At the Shomrom the IDF entered Nablus faster than planned and thus a plan to drop leaflets over that highly populated city was canceled during the morning. Ouragans attacked the retreating Jordanian forces at Farah Valley. Around the Jerusalem area the IDFAF was very active. During the morning the IDF vice commander, Aloof Chaim Bar-Lev, was flown to Jerusalem in a 100 Squadron Piper Cub to supervise the operation at that area. Shortly after his arrival four quartets of IDFAF aircraft, including at least one of 105 Squadron SMBD, gave CAS to the IDF 55th Brigade at Augusta Victoria on the northern outskirts of Jerusalem. Simultaneously the Magisters of 12 FS continued their BAI effort east of Jerusalem with the unit's C.O. Rav Seren Aryeh Ben-Or orbiting over the area and directing the formations to their targets in spite of the fact that he himself exhausted his ammunition. The formation led by Rav Seren Ben-Or claimed the destruction of seven of the 15 tanks that they discovered and he directed another three formations to the spot to continue the work. The magisters claimed the destruction of some 50 armored vehicles and another 70 vehicles east of Jerusalem. Even if somewhat exaggerated, there is no doubt that the contribution made by the armed trainers was indeed remarkable.

During the afternoon, the Jordanians were in full retreat from that area. The IDFAF were attacking the retreating columns, losing a single aircraft in the process, when Segen Ya'acov Zik of 113 Squadron had to eject from his crippled Ouragan just east of the River Jordan, after attacking a Jordanian camp near Allenby Bridge. Other Ouragans immediately covered their comrade while an S-58 was sent to rescue him. The helicopter captain, Rav-Seren Kohav Hess, ignored small calibre hits which caused a highly dangerous fuel leak and rescued the distressed pilot back home. While the conquest of the West Bank was virtually complete, 103 Squadron and its Noratlas twin-boom transports provided vital support to the fast advancing army. One was damaged by AA fire during a fuel-drop mission, while another made the first landing at Jabel Libni air base, where only 1,100 meters of the main runway's 2,800 meters remained intact. Yet one more mission became necessary due to the rapid advance at Sinai. That was an air defence suppression at the Suez Canal Zone in order to allow the supporting IDFAF aircraft the same freedom they enjoyed over Sinai. The only remaining threat were the SA-2 batteries stationed in the area and these were attacked by Vatours of 110 Squadron and Mirages of 101 Squadron, one of the latter being lost.

During Day 3 support operations in the Sinai, the Israelis lost another two aircraft, the first was a Mystère of 116 Squadron which was engaged in the attack of retreating Egyptian forces at the

Part of the success of the IDAF as a fighting force are the highly trained and motivated ground crews, capable of making a very fast turn around, to increase the number of sorties of any given aircraft. (IDF)

Milah Pass, where hundreds of retreating vehicles were caught by the IDFAF, destroying many, while the remaining were abandoned by their occupants. The Mystère pilot was covered for about an hour by other aircraft until relieved by an S-58. The second loss was an SMBD of 105 Squadron of a formation that was intercepted by EAF MiG-17s. Although the French fighter was claimed to be superior to the MiG-17, it should be remembered that the Soviet interceptor had exceptional performance at low level. The Israeli pilot, Segen Yigal Shokhat, ejected. A 100 Squadron Piper that was in the area, and which was in fact responsible for the dogfight when it was intercepted by the MiGs, with only the SMBD arrival saving it, landed to rescue the pilot, but he refused to return in the Piper and elected to wait for the S-58 which was sent to bring him back. The combat was still going on however and three Mirages of 119 Squadron on an interdiction mission to Ras Sudar/Sharm El-Sheikh were ordered by the controller to abandon their original mission and help the SMBDs. This they did with remarkable success as Segen Giora Rom shot down two of the MiG-17s to become the first Israeli jet ace and the leading scorer of the Six Day War. The mission that the three 119 Squadron Mirages had to abort was connected with the airborne assault at Sharm El-Sheikh, which was occupied unopposed.

Day 3 saw only negligible effort by the EAF when compared with the previous day, although a Noratlas flying as a communications relay near Bir Hassaneh at Sinai was attacked by EAF MiGs but managed to evade them. The least active front, as far as the IDFAF was concerned, was the Iraqi one. Although there was no operational need to attack H-3 air base, a third air raid was launched that morning and it cost the IDFAF dearly, reminding its planners that even in a successful campaign only absolutely dangerous targets should be attacked.

Once again, four Vatours of 110 Squadron were the core of the strike force while their escort was provided by a quartet of 117 Squadron Mirages. Once more the attacking force was intercepted over H-3 by IrAF Hunters, but this time with catastrophic results for the attackers. The Hunters shot down two of the Vatours, including the leading Vatour IIN, and a single Mirage. Two pilots, Segen Gideon Dror — who ejected from his Mirage, and Seren Itshak Golan — who ejected from his Vatour, were taken POW by the Iraqis. Seren Shlomo Keren and his navigator Seren Alexander Inbar were killed when their Vatour IIN crashed at Saudi Arabia.

The 117 Squadron Mirages claimed only two of the defending Hunters, with the first credited to Dror. That kill cost his own ejection. While concentrating on the Hunter in front of him, he failed to notice a second Hunter at his six o'clock position. It is interesting to note that the Iraqi pilot was the victim of the same misjudgment as the Israeli leader, Seren Ezra Dotan, saw what was going to happen and shot down the Hunter that risked Dror, but a second too late. All in all, the third and last air raid on H-3 was a failure but the effort at that front did not end.

The fate of the missing aircrews was not yet known and, in case they were still hiding at the desert, three Mirages were sent to orbit over the area trying to pick-up any message that the missing aircrews might transmit by the survival radios. The Mirages failed to locate the missing aircrews, but on their way back home were intercepted by Syrian MiG-21s, which managed in a perfect GCI (Ground Controlled Interception) to sneak undetected behind the Israeli fighters. Fortunately enough, they only managed to damage Seren Dotan's Mirage with an AA-2 Atoll AAM. In spite of extensive damage, Dotan was able to bring his Mirage to a safe emergency landing at Meggido airstrip. By that evening a Noratlas, configured as comms relay, and three Super Frelon SAR helicopters, were alerted to stand by for a possible rescue effort, but as the missing aircrews had not been located by another Noratlas that was sent for that mission near H-3, they were not launched. Although the Noratlas that was sent to H-3 failed to receive any transmission from the downed aircrews, it attracted the attention of the Iraqi defenses that launched SA-2s at the venerable transport. These missed and an attempt to intercept it by a MiG-21 also failed. Similar efforts were conducted on the next two nights but without any apparent success.

By the end of Day 3, the West Bank was captured completely, the IDF forces in the Sinai controlled most of the peninsula and were on their way to the Suez Canal. At the Golan Heights the IDFAF was operating against the Syrians in anticipation to an army offensive. The only front on which hostilities ended during Day 3 was the Iraqi, since no more air raids were launched against H-3 air base. Israeli losses have been heavy, with nine aircraft written-off in action: two Mirages, two Vatours, two SMBDs, two Mystères and one Ouragan, five in air combat and four by AA fire.

THE IDFAF ON JUNE 8

As the IDF Armored Corps were about to reach the Suez Canal, the IDFAF continued its efforts to destroy the SAM batteries defending the Canal Zone; during such an attack on a SA-2 site at Tel al-Kabir, some 30 miles west of the Canal, a 101 Squadron Mirage was lost, with its pilot Seren Romach killed. A special effort, that also continued from Day 3, was made to destroy the vehicles of the retreating Egyptian forces and a major success was achieved at the Mitlah Pass, where hundreds of vehicles were either destroyed by Israeli aircraft or simply abandoned by their occupants. Almost all the IDFAF combat units took part in the blockade operation, including 110 and 117 Squadrons. 105 Squadron was also very active that day at Sinai attacking Egyptian forces scattered at the peninsula. The EAF activity over Sinai was quite weak, but no chances were taken and the Mirages flew many boring CAPs over Sinai. Segem Rom of 119 Squadron, for example, flew two such missions that day without encountering anything hostile in the air. Luckier was Zoric Lev, on an emergency posting to 101 Squadron, that flew a CAP at northern Sinai during which EAF MiG-17s were intercepted over Romani with Lev claiming one of them. Four more EAF fighters were claimed by IDFAF pilots that day including two MiG-19s shot down south of Bardavil, a MiG-21 shot down in that same area, and another MiG-21 shot down further west near the Suez Canal. Seren Menahem Shmul, IAI Chief Test Pilot during the early nineties, raised his personal score to three kills when he shot down an Il-28 and a MiG-21 in a single mission over Sinai, adding to his first kill, a MiG-19 claimed earlier that day. Other EAF aircraft operated at Sinai but were not intercepted. They tried to shoot down an Israel Dakota west of Bir Gafgafa, but the transport evaded the attackers. Another supporting aircraft which saw MiG-19s was a Bell 47 on a resupply mission, but apparently the rotorcraft was not seen by the MiG drivers.

One more CAP formation at Sinai was to see action that day although this action developed into a diplomatic incident with the United States. At 06:00 that morning, a Noratlas flying a maritime patrol mission identified a U.S. Navy intelligence gathering ship north of El-Arish. A tragic mistake led to the incorrect assumption that the vessel had retired from the area five hours later, at 11:00, when it was reported that El-Arish was shelled from the sea, reports that later turned out to be false. A Mirage formation flying CAP over northern Sinai just before noon was ordered by the controller to search for an Egyptian ship that presumably was shelling IDF ground forces at El-Arish. The only ship in the area was the U.S.S. Liberty, on intelligence gathering duties, attached to the U.S.N. Sixth Fleet. The Mirages were authorized to attack the ship, it is claimed, only after the enquiry with the U.S. Embassy resulted in a claim that no U.S. Navy vessels were in the area. Being configured for air combat, the Mirages could only strafe the ship. An SMBD formation was scrambled to the area and Israeli Navy torpedo boats were directed to intercept what was misidentified to be an Egyptian Navy supply ship, named El-Katsir. However, when the two SMBD arrived, but only after dropping their bomb load, one of the pilots was able to read several of the identification code letters, written on the ship's hull, although nobody saw any flag. After transmitting back to the Operations Center what he saw, the leader was ordered to stop the attack since it became obvious that the ship was not Egyptian. This information was not passed, however, to the Israeli Navy and hence its torpedo boats pressed on the attack and launched six torpedoes before identifying the U.S. flag.

By this time the Liberty was severely damaged, 34 of its crew were killed and 171 injured. Once the mistake was realized, the IDF offered to fly the U.S. Air and Naval Attachés to the stricken ship and enquire if any help was needed. The offer was accepted but the Captain of the Liberty did not accept the Super Frelon aboard and the ship sailed under its own power to Malta. Help was provided by the carrier U.S.S. America, which even had a counterstrike force readied on deck, in case it was necessary. Fortunately the crisis was immediately defused.

Up north, missions over Syrian positions continued, a noteworthy attack that day was recorded on the Jalabina strong point by the

Mystères of 109 Squadron. One of the participating pilots in that sortie was Seren Asaf Ben-Nun, shot down over exactly the same spot on the previous day. Other aircraft continued to bomb the Syrian positions, challenged only by the intensive AA fire. A 113 Squadron Ouragan was lost east of Kuneitra. The pilot, Seren Otniel Shamir, lost his life. That day, possibly encouraged by the successful interception of 117 Squadron Mirages returning from H-3 air base the previous day, SAF MiG-21s appeared over the Golan Heights. They were intercepted by 117 Squadron Mirages and Seren Ahud Hankin shot down one of the MiGs but his Mirage was damaged by debris and he had to eject over Ramat David.

By the end of Day 4 it seemed that the war was almost over. The west bank was in Israeli hands, the remains of the Egyptian Army almost completed their retreat from Sinai, and at the north the Syrians were only shelling, not attacking. But, as would become clear the next day, the war was not over yet, only the focus shifted from the Egyptian to the Syrian front. The unbearable situation, with the Syrians on the top of the Golan Heights and freely shelling Israeli civilian targets, was to be drastically changed on Day 5. But on Day 4 the only signs that an IDF offensive at the north was about to be launched was the notification to the helicopter unit's C.O. to prepare for a major heliborne assault at the Golan Heights.

THE IDFAF ON JUNE 9

During Day 5 the activity on the Egyptian front was fading while the IDF assumed control of the peninsula. The important oil fields at Ras Sudar were captured by a heliborne assault when a pair of Super Frelons, led by the unit's C.O., Rav Seren Chaim Nave, landed the first troops, under fire, very close to the Egyptian barracks. Once on the ground, however, the IDF troops met no real opposition and reinforcements were flown in by the following S-58s. The heliborne forces, that were covered by Mirages, occupied the area and a civil Hiller UH-12E-4, used for oil support missions, was captured intact. Since the IDF was present on the east bank of the Suez Canal, it seems that the Egyptians feared that it might try to cross the Canal to the west and, hence, the attack activity of the EAF intensified. IDF forces were attacked by MiG-17s but the EAF lost four of the attackers. Two MiG-17s were claimed by Seren Asher Snir of 119 Squadron that flew CAP at Sinai before intercepting the MiG-17s in the Ismailiya area and continuing the combat until the Nile Delta was reached. The other two MiG-17s were intercepted by a pair of 117 Squadron Mirages that escorted Vatours on an attack mission at the Suez Canal, each of the Mirage pilots claimed a single kill. Sheren Hankin shot his victim fairly quickly and over Sinai, but Seren Lanir had to chase the other MiG-17 well west of the Suez Canal before achieving a successful shooting opportunity.

The focus shifted to the north where the IDF launched a ground assault to occupy the Golan Heights.

Beginning at dawn, the IDFAF intensified its attack effort in the area and even the 12 FS Magisters were called upon to participate once again. This proved a fatal mistake as they were just too slow and vulnerable to the massive AA fire at the Golan Heights. There was no need to risk the Magister pilots that contributed so much to the war effort during the initial stages. This mistake was realized and the Magisters were retired from first line duties, but only after the unit C.O., Rav Seren Aryrh Ben-Or, was lost to flak as the 12 Flight School's sixth and last pilot killed, this time while attacking a Syrian Katyusha battery. Another aircraft lost that day at the Golan Heights due to AA fire was an SMBD flown by Shlomo Ben-On, the former 105 Squadron C.O., and then vicecommander of Wing 4. He was on an emergency posting with the Scorpions. A Dakota was sent to the west of the Golan Heights and rescue helicopters were placed on alert, but Ben-On was not seen to eject from his aircraft. That day 110 Squadron changed its tactic over the Golan Heights, bombing from 15,000 ft. to avoid the intensive flak and diving down to strafe Syrian targets only at the absence of AA fire. That tactic proved its worth as far as losses were concerned and the unit, whose prime target during the three last days of the war was the Golan Heights, did not lose a single Vatour in those attacks.

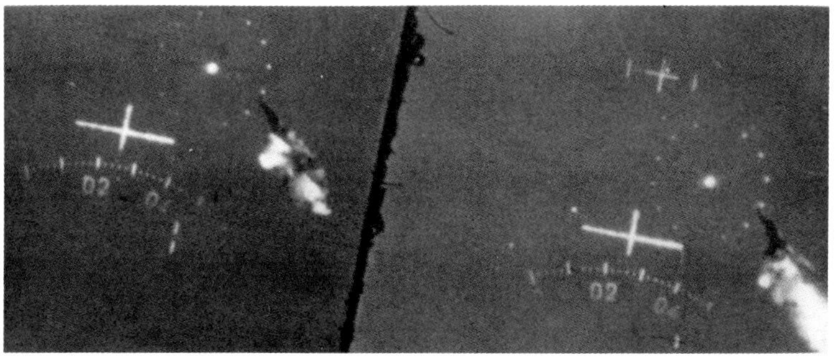

Mirage against Mirage! - This unique gun camera still was from a Mirage or Nesher, while shooting down with gunfire a Libyan Arab Air Force Mirage 5D over the Mediterranean Sea during the later stages of the Yom Kippur War. (IDF)

THE IDFAF ON JUNE 10

During Day 6 fighting continued only at the Golan Heights with the IDFAF flying attack and support missions. Down south at Sinai it concentrated on the support of ground forces and Bir Gafgafa AB (to be known for the next dozen years as Refidim until returned to Egypt) became the center of activity. Noratlas transports established an air bridge carrying fuel, while a Hawk SAM battery was deployed to Refidim to protect the area. Special effort was made by the helicopters that were involved in a major heliborne operation on the southern side of the Golan Heights. Syrian fighters did not challenge these missions, except for a couple of MiG-21s which intercepted a helicopter on a commander transport mission, carrying the IDF CoS R/A Itshak Rabin and the Northern Command C.O., Aloof David Elazar. The helicopter's crew evaded the MiGs until the Mirages that were scrambled to the area removed the threat.

Yet one more mission flown by the IDFAF helicopters at the Golan Height that day was aircrew rescue and the last pilot to abandon his aircraft during the Six Day War was S/A Shmuel Sheffer, the Tel Nof air base C.O. He flew as wingman to 119 Squadron vicecommander when his Mirage was hit by either flak or debris from the explosion of his own bombs. Sheffer ejected very near the column he was attacking but his leader covered him until the pair that was to relieve them on the attack mission arrived.

That pair consisted of two 101 Squadron Mirages led by S/A David Ivry, the 12 FS C.O., on an emergency posting to that squadron. The two pilots that took off at 11:51, as the third 101 Squadron formation sent to the Golan Heights that day (the previous two took off at 04:50 and 08:15) covered S/A Sheffer until the helicopter that was called to retrieve him arrived, as well as two SMBDs that carried on the suppression of Syrian ground defenses until the successful end of the retrieval.

Until noon the IDFAF flew many CAS and attack missions at the Golan Heights, including the destruction by 107 Squadron Ouragans of a Syrian ambush at Vasset, where Syrian SU-100 tank killers concealed in houses fired their guns through the windows. By noon, however, the final effort was over and the cease fire marked the end of the conflict.

Thus the Six Day War ended in a clear Israeli Victory. There is no doubt that the IDFAF was the prime reason for that overwhelming success. Obtaining complete air superiority early in the conflict through Operation Moked allowed the full support of the ground forces which did a magnificent job in occupying Sinai, the West Bank and the Golan Heights. During the Six Days of the War the IDFAF used its resources to the full, with all its units, from the Mirages down to the Pipers, heavily involved in the fighting.

As a consequence, the losses were quite severe, with a total of 46 aircraft written-off and another 23 severely damaged. Since Israeli combat aircraft flew 3,279 sorties, this total (excluding the Noratlas) amounts to a loss rate of 1.37%. In human terms the losses comprised 24 aircrews killed in action and another seven taken prisoners in four Arab countries. The units that suffered the most were 105, 110 and 113 Squadrons as well as 12 Flight School, each losing at least five aircrews either killed or taken POW. Three units managed to end the war without suffering any loss of the like. These were 107, 109 and 119 Squadrons, although 109 lost a pilot as a POW in Lebanon. A summary of known losses (only three aircraft w/o, of which it is presumed that two pilots were killed, are missing from the table) is given:

		AIRCREW FATE			
AIRCRAFT	KNOWN LOSS	RESCUED	POW	KIA	UNKNOWN
Ouragan	9	4	1	4	
Magister	6	6			
Mirage	10	4	1	4	1
Mystère	5		3	1	1
Noratlas	1				
SMBD	7	1	1	5	
Vatour	5	1	3	2	
Total	43	13	7	22	1

As for the Arab losses, there is no doubt that the IDFAF caused a lot of damage to the infrastructure and installations, destroyed hundreds of armored vehicles, and possibly even a higher number of vehicles of other types. Even the claim of 12 FS Magisters is quite impressive, with 128 tanks, 43 armored fighting vehicles, 292 trucks, 46 trains and 47 AA guns as artillery pieces. To this we must add the claims of other squadron and, if a bit exaggerated, they are impressive and not far removed from the true number as shown by the many vehicles found after the war at the occupied areas. But above all, the Six Day War will be remembered for the almost complete destruction of three Arab air forces: the EAF, SAF and RJAF. All in all, the IDFAF claimed the destruction of 452 aircraft, of which 58 were destroyed in the air. Two were seen by IDFAF aircrews to crash while engaged in air combats, two were shot down by AA fire and the rest were destroyed on the ground. The actual split of types is given in the two tables below:

AIR FORCE	MiG-21	MiG-19	MiG-15/17	Su-7	Hunter	Il-18	Tu-16
EAF	100	29	89	14		29	30
IrAF	15				5		1
LAF					1		
RJAF					21		
SAF	33		23			2	
Total	148	29	112	14	27	31	31

AIR FORCE	Mi-4	Mi-6	An-12	Il-14	other helos	Unident
EAF	1	10	8	24		4
IrAF				2		
LAF						
RJAF					2	6
SAF	3					
Total	4	10	8	26	2	10

Even if it is assumed that this claim is a bit exaggerated, there is no doubt that the RJAF was practically wiped out, the EAF got very near that point, the SAF suffered heavy losses and the IrAF got very near that point, while the LAF loss was merely an episode. No matter that the exact amount of damage caused to the Arab air forces is debatable, the fact that it was inflicted to win air superiority must not be overlooked; that was the major test of Operation Moked, not the exact number of destroyed aircraft. In the light of this test, the success was tremendous as complete air superiority was indeed achieved. It was not anticipated, however, that, following Operation Moked, the Arab air forces would simply vanish and the fact that a portion of their force continued to operate until the very end should not cause any surprise. Since small scale operations continued, air combat followed the same trend, in addition to those who fought during Operation Moked with the IDFAF, claiming the destruction of 58 enemy aircraft (downgraded from the initial 77 claims) in the air, including 48 by Mirages, five by SMBDs, three by Mystères and one each by Vatours and Ouragans. A daily representation is given in the text but obviously a few kills went unrecorded and hence a summary is provided in the following table.

	MiG-21	MiG-19	MiG-17	Su-7	Hunter	Il-14	Il-28
Mirage	15	12	9	5	5	1	1
Mystère			2			1	
Ouragan	1						
SMBD	2		1		2		
Vatour						1	
Total	18	12	12	5	7	3	1

Against these claims the IDFAF lost in air combat a total of eleven aircraft, plus a twelfth that was lost to fuel starvation during its return from a dogfight. Hence, a favorable exchange ration of five enemy aircraft shot down to each Israeli aircraft lost in air combat was achieved. Of the eleven losses, four were Mirages: two to EAF MiG-21s, one to an EAF Su-7 and the fourth to an IrAF Hunter. Three were Super Mystères: two to EAF MiG-21s and one to an EAF MiG-17. A single Mystère was downed by an RJAF Hunter, and three Vatours were downed, two to IrAF Hunters and one to an EAF MiG-21.

The leading IDFAF ace was Segen Giora Rom of 119 Squadron and, interesting enough, all three pilots that came in second place with three kills each were also of 119 Squadron — these were Seren Oded Sagee, Seren Menahem Shmul and Seren Asher Snir. Those four scored a total of fourteen kills, two more than did the entire Mirage equipped 117 Squadron that claimed a total of twelve. The total claim of 119 Squadron was 21 kills, of which 18 were scored by six pilots. 101 Squadron claimed 15 kills. Among the pilots with two kills to their credit were 119 Squadron C.O. Rav Seren Ran Peker, his deputy Seren Eitan Carmy and Seren Yehuda Coren and Seren Ahud Hankin of 117 Squadron.

To summarize, the Six Day War was a magnificent achievement. The Mirage established its fame worldwide; however, harder times were ahead. During the War of Attrition and the Yom Kippur War the Mirage, together with the native developed Nesher, established new landmarks in the story of air warfare.

MIRAGE PILOTS' NARRATIVES - SIX DAY WAR

Segen Giora Rom, from 119 Squadron, was the IDFAF's only ace during the Six Day War, and the first Israeli jet ace. He bagged two Egyptian and one Syrian MiG-21s on June 5, and two Egyptian MiG-17s on June 7. Here is his personal tale of the five kills:

". . . The scene: The first day of the Six Day War, June 5, 1967. At 09:00 hours I was scrambled, forming part of a Mirage formation, the briefed target was Egyptian airfield at Abue Sueir, west of Ismailia. It was my first sortie into enemy territory. I was a Mirage pilot, a Segen (Lieutenant), and at 22 years old one of the youngest in the squadron. As our previous mission had been local air defence, we had not enough time before scrambling to change to longer range external tanks and thus were a little tight on fuel. We climbed to gain the briefed altitude and set course for Egypt. As we reached the target area, we noticed that the air base had already been visited. Runways were a shambles and some Il-28s were burning. Suddenly as we came in, an Egyptian MiG-21 started its take-off run on the parallel runway. I went in low enough and manoeuvered into attack position behind him.

Mirage IIIBJ number 787 in the early markings of 119 "Bat" Squadron with the natural metal finish. (IDF)

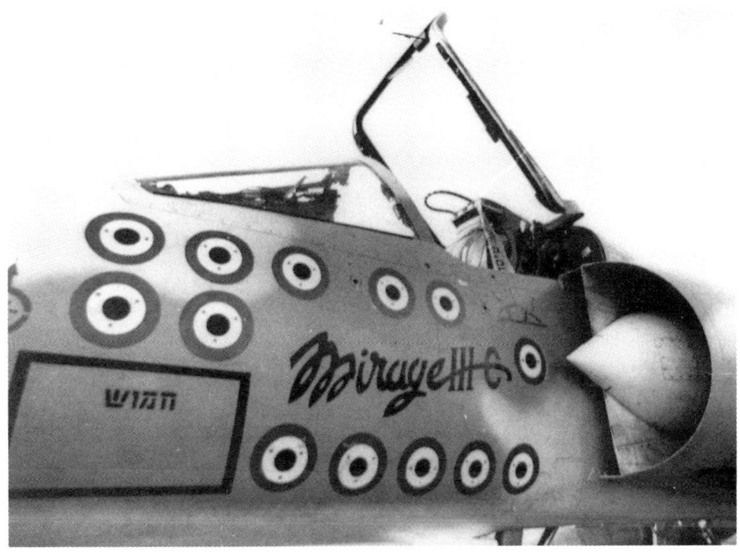

Number 159 was the top scoring Mirage airframe with 13 confirmed kills, both Egyptian and Syrian aircraft, the photograph was taken shortly after the October 1973 War while in service with 101 Squadron, note the black/ yellow identification triangles. (IDF)

My number one stopped me: — "Don't touch him! I'll shoot him down." — I moved aside and the leader went in. That was that. It was the first time I actually saw an enemy aircraft explode. As if by magic, the whole sky filled with aircraft and all hell broke loose! Explosions all round! Dogfights all over! Smoke and fire! Through the haze I saw two MiGs chasing a pair of our Mirages. I went into afterburner and closed in on one of the MiGs. I shot at him from 200 meters range and missed. I felt terrible. What a rookie! All my self confidence seemed to fade away. For three years I had prepared myself for this moment and when it came I had failed!. I tried another go. Closing to 150 meters this time I fired another short burst. The MiG exploded into a searing ball of fire, right before my eyes. I felt nothing.

On my way home, I caught another MiG-21 and after a chase shot him down. I was drunk with elation. Now I was short on fuel and hurried for home. During the flight I heard the leader report he was bailing out due to fuel exhaustion. I reached base on my last drops. My first sortie and two down!

During the afternoon I was briefed to attack Cairo International. While sitting strapped in my cockpit, a messenger came up and handed me a map with new instructions. The target: T-4 in Syria, one of the farthest airfields in that country.

We took off in a four ship formation, following our leader Eitan Carmi. The flight was long, about thirty minutes in each direction. We reached the target and saw two MiGs patrolling over their base. Sitting ducks. But our orders were strict. No deviation from the plan.

As we circled the airfield, Asher Snir downed one of the MiGs. We others went in strafing. As I came out from my run, I saw a MiG some 1,000 meters away. Warning the others, I dropped my fuel tanks and turned into him, who started to take evasive action. As we tangled with each other, several other Mirages arrived over the field. One of them pulled up to fire on "My" MiG and stalled. The Syrian stopped minding me and my attention turned to the stalling Mirage. I was right after him, the calls of the stricken Mirage pilot filling my headset. The MiG launched two missiles towards the Mirage, but missed both times. I aimed and fired. He caught fire over the desert airfield and crashed in flames. I was alone. Climbing up to 35,000 ft. I crossed the Syrian border, short of fuel. I saw Lake Galilee and felt much better — home was near.

Next day I flew with a formation in an attack on the Golan Heights. We were to strafe a Syrian tank position overlooking the Jordan River bridge. The entire Hula Valley below seemed on fire. From the air everything was covered by thick smoke. We went in low to strafe the tanks and encountered strong flak. During my third pass, I felt a tremendous crash below my seat. — "That's it" — I though. Checking the instrument panel, I could not see any lamps lit. All systems were dead. I knew that I would have to eject and turned sharply to cross the Jordan to the west. The gauge showed 10,000 rpm. Up to this moment I had known that 8,700 rpm was the maximum before seizing-up. My engine had really gone for a burton, but still I flew, there was no electrical power to the controls, the stick became rigid and stone-heavy. I grabbed it with both hands, but its tendency to go alone forward was difficult to stop. Frantically I tried to climb, but the systems wouldn't respond. I was on course for base at Ramat David and suddenly found myself above it. Making a final approach with a dead aircraft, I hopefully dropped the landing gear, but no indicators were working. I tried to guide the airplane away from the sun, trying to see my own shadow and whether the wheels were fully deployed. I could not line-up with the runway, the aircraft was hardly functioning. I flew off to the west in a wide turn until I aligned with the runway and sank slowly, hanging on to that shaking stick with both hands, praying. I just made it down and stopped the Mirage. I could hardly move. Afterwards a check revealed that a 37mm round had hit under my seat, damaging the left side of the cockpit and destroying the control lines. There was some shrapnel in my leg, which I had not even felt during the flight!

I spent that night in the hospital, but I could not rest. Here I was on my back with the war running away from me. I talked my way out and back to the action.

I was back in the flying roster of my squadron next day, assigned to attack an Egyptian armored column near Sharm El-Sheikh. On our way, the controller ordered us to drop our bombs and change to a new heading at 20,000 ft. I caught a glimpse of a MiG flying close to the ground which I lost almost instantly. I dropped my fuel tanks and dived, calling the others to follow me down. As we closed, I saw two MiG-17s fighting two Super Mystères. I entered into the battle, flipping upside down to mark where one of our planes was going down. The two MiGs were down on the deck by now, racing home. I closed in fast, leaving my two wingmen behind. Those two MiGs were mine alone! I closed range on the first and let fly with my guns. I glimpsed part of him flying off and he steered, climbing for his life. The other broke contact and disappeared.

I stayed on the first MiG, and was so close that I could make out the pilot and the instrument panel. Suddenly he ejected and the 17 smashed into the ground below. I rolled out and started to chase the other MiG. Flying very low to silhouette him against the horizon. I was blinded by the setting sun. I found the MiG over the swamps north of Ismailia and we started to dogfight over the Canal. The man was stubborn and tried all the tricks in the book. I was hard at it, trying to hold on him, when I received an urgent call from the formation leader to breaking contact and return east, but before turning for home, I connected a solid burst into his left wing root, and the left wing parted company with the rest of the aircraft which fell away and smashed into the ground. No ejection was seen . . ."

Sgan Allof (Lt. Col.) David Ivri <in 1995 an Allof (General) in the reserve> tells his own account in the following lines, but before that it may be interesting to take a quick glance at his career in the IDFAF. Born in 1934, he joined the air force in 1952 and graduated as a pilot in June 1954. After converting to fighters with 105 Squadron, then an Operational Training Unit equipped with Spitfires, he flew with 101 Sqn.(Mustang), 117 Sqn.(Meteor), and 113 Sqn. (Ouragan). In 1956 he was sent to a gunnery course in the RAF. During the Sinai Campaign he flew Ouragans, later he served in the Flying School, in Mystère (101) and Super Mystère (105) squadrons, graduating from the Command and Staff School in 1960. 117 Squadron vicecommander (Mirage), 109 Sqn. (Mystère) C.O., 117 C.O. and Flying School C.O. until 1968 when he was appointed head of the Operations Department. In July 1970 he became Tel Nof air base C.O. and during 1973-1975 he was head of Air Wing. In July 1977 he graduated from Technion as an Aeronautical Engineer and on October 28, 1977 was appointed as the IDFAF C.O., a post he occupied until December 1983. One of his latest posts has been as the general director of the MoD. During 1987, his son, Seren Gil Ivry, a fighter pilot in the IDFAF, was killed in a flying accident.

". . . Our mission was an attack on the Golan Heights on the morning of Saturday, June 10. During that day, the IDFAF concentrated almost exclusively against Syria. Only a small number of sorties were flown to other targets. The rate of sorties was slowed down after fighting had died down in the Egyptian and Jordanian fronts and air superiority was achieved in Syria.

It was only several weeks after my appointment as Flying School commander. Previously I was a Mirage squadron C.O. at Ramat David (117 Sqn.). During the war I was posted to another Mirage squadron (101) as a flight leader and reserve C.O. The combat days in my new squadron mixed me with that unit and distanced me from my old squadron. I understood the meaning of war. Common experiences in a war are much more dominant than any other tie. The war creates an emotional link between the warriors, a kind of 'warriorhood' and that is what this story is about.

My number two was Giora Epstein, then a junior Mirage pilot and a former paratrooper. We had flown some sorties together during the war and the short period of tension before it. I was very impressed with him. He was a pilot that could be trusted. Later he commanded 117 Squadron, performed many operational sorties and scored 17 kills, thus becoming Israel's Top Gun. Our aircraft were armed with two 500-kg bombs and 30mm ammunition for its two cannon. We strapped, turned on the radio and waited for the scramble signal. In order not to miss our turn we informed of our existence on the radio every few minutes. An unnecessary action but one that strengthened our confidence. It was almost noon and the long time in the cockpit started bothering me. I was afraid that, because of the pettiness about us, we would be replaced and minutes later, when our replacements occupied our places, there would be an scramble. We would pay the price of waiting and our replacement would win the scramble. At least we were scrambled. The mechanics got into action, engine rpm up, throttle to neutral, engine start, checks . . . The Mirage has a soul, it is not just an aircraft.

In the air we hear that another formation of another Mirage squadron (119), had been sent before us on the same mission. Our turn would arrive when they finished. Our route crossed the Shomrom area, only days before a forbidden place for overflying. The plan is to cross the Jordan River south to Lake Galilee and enter the Golan Heights from the south and not from the west.

We listen to the other formation and it is clear that something has happened. The leader is calling his number two, but receives no answer. We all get to radio silence in order not to miss an answer, if it would come. After a short while, the leader reports in high tone that his number two is on fire and probably ejected. We wait in silence and then he reports that he sees a parachute. We are southeast of Lake Galilee and Giora informs that he had a visual on the parachute. I order him to preserve "tally-ho" and lead me until I could see it by myself. I was told about Giora, with his "hawkeyes," that he sees aircraft beyond the usual range, and here the guy has proved it. From a range of 12-15 km, and looking down, he located the parachute, I located it at a range of 3-4 km!

The leader was much relieved when he heard that we are on our way, he was running out of fuel. It was decided that my formation would take responsibility for the rescue. We figured that it would take him five minutes to reach the ground and we were there before he touched down. While waiting for him to land, we heard his leader talking with the controller. He was bothered and hinted that the pilot that ejected is more than a common pilot.

Later we discovered that he was Sgan Allof Sheffer, the Tel Nof base commander.

The landing point was in the Sheih Maskin area near Syrian forces. They fired but we didn't know if at us or at the pilot hanging in his parachute. We started an attack run upon these forces in order to keep their heads down. I informed Giora that we would perform passes in singles, use minimum of armament and preserve fuel for long endurance. We descended to 500 feet and located a BTR and many other vehicles near a road. They all fired and it was time to start bombing. Upon completing my first pass, I saw the pilot running towards a rocky hill, only later we knew that he had broken a hand and a foot on landing, yet he was running like hell. The distance between the hill and the road was one kilometer. On the road stood large Syrian forces, we were running against time. The questions were: 'When would the rescue helicopter arrive? Would we run out of ammunition? Would we have enough fuel? When would the Syrians start to move toward the hill?'

Giora also understood the problems and we had excellent coordination between us. We fired short bursts along the Syrian troops. We were efficient and the Syrians hadn't started moving yet. After we completed the bombing runs, we started strafing passes. On the radio we hear that the helicopter is on its way. We gave him directions and other data and hoped to lead him exactly to the pilot. suddenly Giora struck again with his "hawkeyes": he located the helicopter from quite a distance and directed it to the hill while I continued the strafing runs. In order to save ammunition for the critical rescue stage, when the helicopter would land, I didn't fire in every pass I made. The helicopter seemed to fly horribly slow. I kept my mouth shut in order not to tell him to fly faster. We were in a critical combat against time. I checked fuel with Giora, he had 300 liters less than I had. In fact he had the minimum fuel to regain home base. He tells me that he is afraid that he has a possible fuel leak. Later we discovered that he was hit by light weapons in the wings and really had a small, but continuous fuel leakage. I figured it quickly. Epstein's fuel situation means he must return, but the situation dictates that he must stay. A moment of thought and then we agreed that he would stay until the end of the critical stage and once the helicopter would take-off, he would clear the area and land at Ramat David.

The helicopter got closer and the Syrians are shooting much more than they did: either because of the helicopter or they just got used to our passes. Anyway we made more passes as the helicopter landed. I passed very low near the landing site and fired my last rounds. I was out of ammunition and so was Giora, he informed me of only 250 liters which could barely be enough. We plotted. He flew to Ramat David and I stayed with the helicopter. I asked the controller to watch for him and give high priority for his landing. Inside I felt that Giora would be OK.

My fuel gauge showed 600 liters and I made one more run to take the fire away from the helicopter. Its the only weapon left. During the pass I observed that the helicopter was near Syrian forces, I warned him on the radio and he changed course. Finally the downed pilot was retrieved, the helicopter's captain informed me that he was injured but otherwise well. I must leave. A pair of SMBDs relieve me in providing escort to the helicopter, and on my way home I hear that the helicopter is already within the area controlled by our forces.

I ask about Giora Epstein's situation. He had landed at Ramat David. I ask for a straight approach and land with 150 liters of fuel. My satisfaction is complete. We recovered a pilot in the middle of the day from the heart of the Syrian army and we all returned. In my heart I compare it to taking away a meal from the mouth of a hungry lion. And what is more important is the fact that the meal was an old buddy . . ."

4

Israeli Mirages' Finest Hour, 1967-1973

The timespan from July 1967 to October 1973 was full of air combat between Israeli and Arab aircraft, especially during the Attrition War in the Suez Canal from March 1969 to August 1970. Up to then it was the longest of the Middle East conflicts and probably came very close to becoming an international conflict when the Soviet Air Forces took an active part in the confrontation, mainly during the fighting in the Canal area.

Until the arrival of the first F-4E Phantoms in September 1969, the Mirages played the prominent role, especially in the air-to-air arena. This period was known in the Chel Ha'avir as the finest hour for its fighter pilots.

But the Cyrano Ibis radar of the Mirage IIICJ continued to cause trouble; it rarely worked at 100 percent. After the Six Day War the Chel Ha'Avir made the decision to remove it from most of the Mirage fleet and a counterweight was fitted in the empty radar dome, although a few retained the Cyrano pegged only in the rangefinding mode. Also the hydraulic system (Hyd. 1 and Hyd. 2) produced early trouble, the warning lights blinking during flights, while on a few Mirages the nose landing gear collapsed during taxiing. Additionally, the Atar 9B afterburner's clamshell nozzles provided some troubles, following an erratic behavior, with one opening and the other closing. Experience with the type progressively solved these problems, aided by IAI, which since the mid-sixties had started to produce spare parts and components for the Mirage fleet, while Beit Shemesh (IAI's engine division) was building some Atar 9C components prior to full production, which was reached in 1968 as part of the trade-off for the ill-fated Mirage 5J buy.

In fact, the surviving Mirages were progressively upgraded by IAI in a program which lasted until the mid-seventies. Apart from replacing the Atar 9B with the 9C, improved avionics were fitted (including new VHF radios, radar warning receivers, and additional store stations), while the wing's main spars and skin was changed, some receiving a grey scheme, others received under fuselage strakes for better controllability at high-angle-of-attack maneuvers.

First air combats after the Six Day War took place on 8 July 1967, when two Mirages intercepted four MiG-21s in the Port Said area. One of the Egyptian interceptors was claimed. A week later on 15 July 1967, the Shafrir I AAM was credited with its first confirmed kill: Rav Seren Ran Peker led four Mirages to intercept eight Fishbeds in the Suez Canal area and shot down a MiG-21 with a direct Shafrir hit from a range of about 1,250 meters over the Nile Delta, as the combat had derived toward the west, also another two MiG-21s were claimed. Later that day, a Sukhoi Su-7 and a MiG-17 were also shot down, while on the debit side a Mirage was shot down by Egyptian AA fire. The Su-7 pilot ejected and landed in the Sinai, being taken prisoner by the Israeli Army.

Three months later, on 10 October, an Egyptian MiG-21 was shot down by DEFA 30mm gunfire. On the night of October 21st, the Mirages helped in the search for survivors of the Israeli Navy destroyer Eilat, sunk off the Sinai coast near Romani by Styx SSMs launched from an Egyptian Khomar class fast patrol boat. The Mirages provided top cover and launched flares to light the area for the rescue helicopters.

During that time, the Soviets provided massive aid to the depleted air forces of Egypt and Syria, while its pilots learned new tactics, trying to master the aggressive Israeli pilots. This reorganization was reflected in the engagements, as until 10 December 1968 there were no conclusive combats. On that day a pair of fighter-bomber MiG-17s were intercepted south of Sharm

Mirage pilot David Ivry, one of the most experienced fighter jocks in the IDFAF, he is actually a General in the Reserve. (IDFAF)

The Israeli developed and produced Shafrir is a larger AAM than the Sidewinder, and has a larger warhead. Early variants were quite unreliable, but the first truly operational was the Shafrir II, employed successfully during the Yom Kippur War by Mirages and Neshers. (IDF)

el-Sheikh, and one was hit by Mirage cannon fire, crashing into the sea. Since late 1967, a four Mirage detachment was forward based at Refidim air base in Sinai for quick interception duties in order to catch Egyptian low level marauders.

On 12 February 1969, in the relatively quiet northern front, a Syrian MiG-21 was shot down with cannon fire near Kuneitra. Twelve days later, the Chel Ha'Avir undertook a massive attack on PLO targets in Syria. Mirages were used only for top cover, bagging two MiG-17s that day, which is thought were flown by Palestinian pilots. On 3 March 1969 a Mirage from 101 Squadron downed an Egyptian MiG-21 over the Great Bitter Lake.

The then Egyptian President, Gemal Abdel Nasser, thinking that the Arab nation honor had been mutilated by the Israelis, refused any type of settlement or diplomatic moves to restore peace to the region. By early 1969 he felt strong enough to try to regain the Sinai by military action. Thus started the conflict which was known as the Attrition War, and which for all purposes began on 8 March 1969 when Egyptian artillery on the West Bank of the Suez Canal started a coordinated bombardment of Israeli positions on the other side, while air activity sharply increased over this area.

The Mirages were used for interceptor, "planned air combats," reconnaissance and top cover for attacking aircraft. On that day, four Egyptian MiG-21s were intercepted and one was shot down over Sinai, the pilot ejected and was taken prisoner. On 14 April, four EAF MiG-21s were intercepted at low level over Sinai by two Mirages, the Egyptians turned toward their own territory and the Israeli pilots chased them. One of the Fishbeds exploded over Egyptian soil, the pilot was seen to eject. On May 5 a MiG-21 was claimed by 101 Squadron. On May 21, again four MiG-21s were engaged, this time by four Mirages, and two of the MiGs were shot down over Egyptian territory. For one of the Israeli pilots, which had blown the left wing of his Fishbed with cannon fire, this was his fiftieth kill.

On 29 May, the Syrians tried their hand at challenging the Chel Ha'Avir but without success. One of their MiG-21s was downed with a Shafrir AAM over the Kuneitra area, its pilot ejecting within Syrian controlled territory.

On June 17, four Mirages on a reconnaissance mission penetrated deep into Egypt, broke the sound barrier over Cairo at very low altitude, and in the process smashed windows all over the capital. The foursome were led by Amos Amir. Nasser's reaction to this operation, which during several minutes nearly scared to death many tens of thousands of Egyptians, was to fire instantly the Egyptian Air Force and the Air Defence Command Chief's of Staff.

Operation "Rimonim" also started in June, an operation in which Mirages flew into Egyptian territory south of Suez City (that location was selected because it was clear of SAM batteries and for this reason was called "Texas" by Israeli pilots). It must be noted that, at the time, the Egyptians had established a formidable SAM array along the Suez Canal. The operation's purpose was to try to draw Egyptian MiG-21s into combat. Nine were downed by Mirages during this operation as follows: a single MiG-21 was downed on June 24 with guns, only a few 30mm shells were necessary and the MiG's rear exploded in a spectacular fashion. On June 26 four Fishbeds were claimed in three separate combats, two with 30mm cannon and two with AAMs, two of the Egyptian pilots were seen to eject. The seventh kill was obtained on July 2, while the last two were a pair of MiG-21s which also were destroyed with guns.

It was very clear that the Chel Ha'Avir enjoyed total air superiority. This was understood by the Egyptians and, under those circumstances, a crossover of the Canal was an invitation to disaster, so they tried other means. July had seen a major escalation in the war with the Chel Ha'Avir being used as "flying artillery" in the Canal Area in a fast reaction basis. When Egyptian artillery opened fire, the Chel Ha'Avir attacked the Egyptian batteries.

On July 8, a major air battle took place over the Syrian border when seven SAF MiG-21s were shot down in two engagements, three and four respectively, all crashing within Syria.

On 20 July 1969, Operation Boxer was started by the Chel Ha'Avir, consisting of attacks over Egyptian positions along the axis from Kantara to Port Said. The EAF rose to the challenge and several dogfights took place in which two MiG-21s, two MiG-17s and an Su-7 were downed by Mirages, while a 101 Squadron machine was shot down by an Atoll AAM fired by a MiG-21. The Israeli pilot ejected over Sinai. Four days later, seven Egyptian aircraft were claimed in air combat, by AA fire and Hawk SAMS: four Su-7s, two MiG-17s and one MiG-21. But that journey was significant because one of the kills was achieved by a

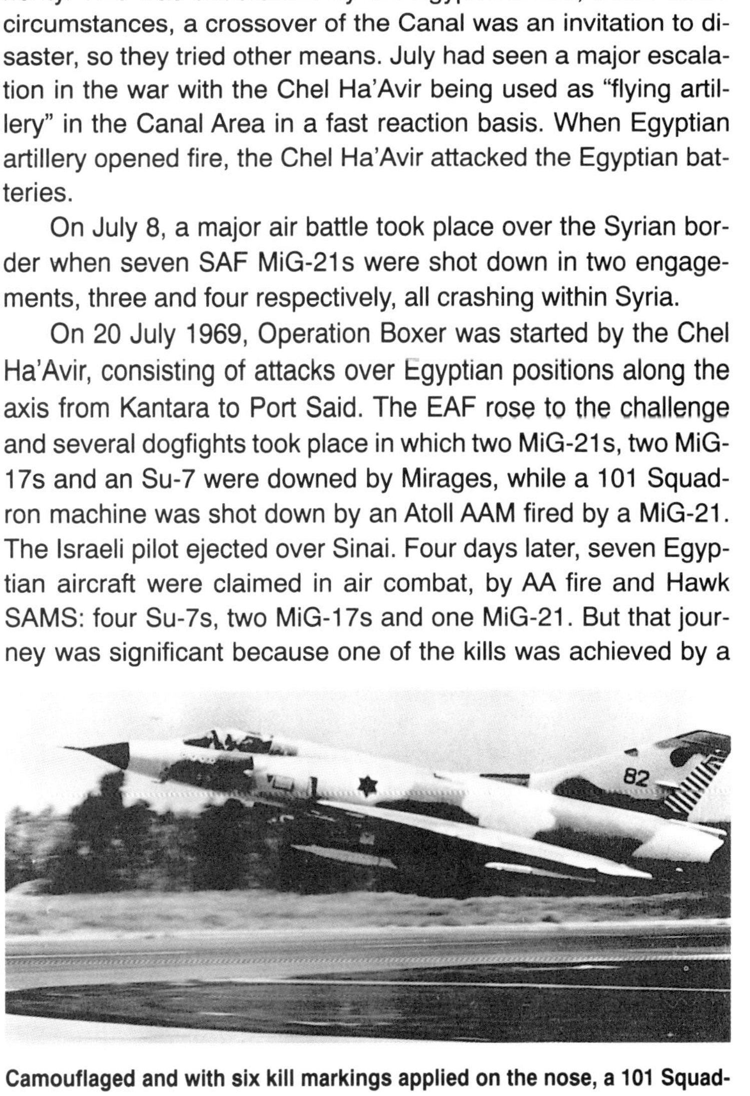

Camouflaged and with six kill markings applied on the nose, a 101 Squadron Mirage IIICJ takes off for a training sortie (IDFAF)

Shafrir II AAM (the pilot was Shlomo Weintraub), which was in fact the first reliable air-to-air missile to enter service with the Chel Ha'Avir, being credited up to the eve of the Yom Kippur War with 13 confirmed kills.

The next Egyptian Air Force effort was on September 11, during the hot morning of that summer day, their aircraft — MiG-17s or Su-7s with MiG-21s as top cover — tried to attack at the Israeli controlled northern part of the Canal Zone. One MiG-21 was downed in combat with Mirages over the Kantara area. Later the Egyptians turned their attention to the Ras Massala area where a MiG-21 and a Su-7 were downed. The Chel Ha'Avir reprisal was not late in coming and during the afternoon attacks were made against Egyptian positions near the Nile Delta. The escorting Mirages engaged in combat with defending MiG-21s, three of which were shot down, while a MiG-21 pilot achieved an clear cut success, shooting down the Mirage flown by Seren Giora Rom (the Israeli top scorer of the Six-Day War) with an Atoll AAM. Rom ejected over Mansura and was taken prisoner. In another air battle on October 6, two Egyptian MiG-21s were downed, both with 30mm cannon fire. Their pilots ejected and landed within Egyptian territory.

On November 11 the first kill was achieved by an F-4E Phantom (on September 5, the first four Phantoms had landed at Hatzor air base and this outstanding "heavy duty" fighter signified a new era for the Chel Ha'Avir). On that day three MiG-21s were downed near Suez City, two by Mirages and one by Phantoms. Note that as sometimes the combats were of a confused nature, from now on perhaps some of the victories credited to the Mirages were in fact achieved by the F-4Es or vice versa. On November 27, a single Egyptian MiG-21 was added to the increasing Mirage tally in an air combat that started in the Ismailia area and concluded with the EAF pilot bailing out about 40 km west from the Canal.

For a change, attention was turned to the northern border with Syria and Mirages bagged two MIG-17s and a MiG-21 near Damascus; incidentally, one of the MiG-17s was credited to Menachem Shmul, who years later — as IAI Chief Test Pilot — made the first flight in the ill-fated Lavi.

The first air combat of the new year took place on 4 January 1970 in the Canal Zone and two EAF MiG-21s were claimed by Mirages, while four days later three Syrian Air Force Fishbeds were shot down in the Seikh Maskin area. On the second day of February, four Mirages led by Ori Aven-Snir, then C.O. of a Mirage squadron, were sent to intercept Syrian MiGs, but one of the Mirages exploded after a lucky direct hit from a heavy-calibre anti-aircraft shell. The Mirage pilot Shlomo Weintraub was killed instantly. The preplanned ambush was abandoned.

On February 8, Egyptian MiG-21s tried to intercept Israeli Phantoms on a deep penetration strike. Two Fishbeds were shot down in the Nile Delta area, at least one with guns by a Mirage. The next day in another air battle over the Canal, a MiG-21 was shot down, but a Mirage from 119 Squadron was lost, downed by a MiG-21-fired Atoll missile. The pilot, Segen Avi Keldes, ejected and was taken prisoner. On February 26, Egyptian MiG-21s tried again to intercept Israeli aircraft on one of their deep penetration strikes. This time a big dogfight developed with the escorting Mirages. Three MiG-21s being shot down, all with guns.

During the month of March, seven Egyptian MiG-21s were shot down by Mirages as follows: two with guns on March 6; a singleton also with guns on March 16; and eleven days later five. During the month of April, action over the Syrian border in the Golan Heights enlivened again and on the second day of that month, three MiG-21s were claimed in air combat, as well as an Hatzor-based F-4E Phantom. On 25 April 1979 Israeli Mirage pilot Amos Lanir claimed an unusual kill when he bagged an EAF Il-28 bomber at night off Port Said with an AAM. On May 12, Mirage pilot Asher Snir downed with an AIM-9D Sidewinder a Syrian MiG-17 over the Mt. Hermon area.

AMBUSHING THE RUSSIANS

The complete air superiority enjoyed by the Chel Ha'Avir, and the deep penetration radius by the new F-4E Phantoms against military targets deep into Egypt, forced Nasser to ask his Soviet friends for more aid. On January 24 he made a quick visit to Moscow and obtained what he wanted. A month later a Soviet Air Defence Division was moved by air and sea to Egyptian bases. Some political and military observers were surprised by the rapid Soviet response, but their reasoning was not difficult to understand: simply, they wanted a foothold in the strategic eastern Mediterranean.

First to arrive was a SAM system with the latest variants of the SA 2 and SA 3 missiles. The first batteries were ready on

Overhead view of a Mirage IIICJ clearly showing the fuselage's area rule concept and the wing delta platform. (IDF)

Mirage IIICJ photographed after the Yom Kippur War with eleven kills, only one of which is Syrian. (IDFAF)

March 15 and deployed to defend Cairo and Alexandria as well as the gigantic Aswan hydroelectric complex, which incidentally had been financed by the Russians. Meanwhile in Israel, Allof Mordechai Hod, the Chel Ha'Avir C.O. and a hard liner, had the idea to attack these air defence arrays immediately. He was told by Moshe Dayan, Minister of Defence, that there was no need — at the moment — for any preemptive action against the SAM sites, as this could cause problems for the state of Israel. In the wake of this controversy, the Chel Ha'Avir abandoned the deep penetration air raids carried out by the Phantoms, nicknamed "blossom" by the crews, and instead concentrated on attacking Egyptian targets on the west bank of the Canal.

During April, the Soviet fighters arrived, comprising about two Regiments with mixed material, with at least 75 of the latest MiG-21MFs, 16 MiG-25s and roughly the same number of Su-15s. This force was deployed to Beni Suef, Mansura, Kom Awshim and Khatamiya, providing cover to the Cairo area and its approaches, with its highly valued military complexes. To protected Alexandria and its very important naval base, a detachment was deployed to nearby Giancallis air base. All the Soviet aircraft wore EAF camouflage and markings for obvious reasons, while all Air Defence Division personnel were Soviet, being under the command of General Vasily Okoniev, formerly the commanding officer of the Moscow Air Defence Division. The first Israeli-Soviet encounter took place on April 18, when MiG-21s tried to intercept a pair of Phantoms on a recce sortie. The Israeli pilots had orders to avoid combat with the Soviet-flown fighters (they knew who they were by monitoring radio traffic). The F-4s made good their escape and returned to their base.

But the Chel Ha'Avir saw the Canal are as a "free-fire" zone to attack in response to Egyptian artillery barrages and the sporadic low level raids by the EAF, whose fighters tried to defend this zone, comprising an area about 25 to 40km west of the Canal, while trying some offensive sorties at the same time. On May 14, two MiG-21s were downed. The next day another of the same type was splashed by Mirages, in both instances after trying to intercept attacking Skyhawks. From May 15, the EAF launched several attack sorties against Israeli positions in Sinai; these attacks caused little damage but — in spite of losses — were intended to be a valuable training aid for Egyptian pilots. That day two raiding MiG-17s were shot down and a further pair the next day, all in air combat and most, if not all, by Mirages.

The next engagement was on June 3, the Chel Ha'Avir attacking the West Bank of the Suez Canal and, as usual, Egyptian MiG-21s rose to intercept them. The MiGs were bounced by the Mirage top cover and three of them were dispatched without loss by the Israelis.

By late June, the Soviets had moved some of their early warning radars toward the Canal Zone, thus providing better reaction time for the SAMs and interceptors. Indeed SA 2 and SA 3 had

Later in its life Mirage IIIBJ 787 was renumbered as 287 and upgraded with the Atar 9C turbojet, as can be seen in this picture, including the identification triangles and camouflage scheme. (author collection)

Two camouflaged Mirage IIICJs on patrol over the Mediterranean, the aircraft in foreground, number 771, has the BU radar warning receiver. (IDF)

been detected near Kathamiya and on June 30 the Chel Ha'Avir finally got the opportune authorization to strike at these sites. The attack was partially successful and two Phantoms were lost. Three of its crewmembers were captured and the fourth rescued by a CH-53 helicopter. In July the Israelis continued attacking these sites and the early warning radars near the west bank of the Canal. As the Soviet system moved east, the air combats decreased sharply, although there were some engagements: on July 10, three EAF MiG-21s were shot down, at least one by a Mirage with guns; on July 27 two MiG-17s attacked Israeli positions in Sinai but, during their egress, were jumped by Mirages. After a breathtaking low level chase, both were shot down over Egyptian territory.

Meanwhile, it seemed that a ceasefire was near after U.S. diplomatic mediation and no serious encounters with Soviet flown MiG-21s had been recorded until July 23. On July 23, two Soviet MiG-21s tried to pounce on A-4 Skyhawks attacking targets near Suez City. The Skyhawks dropped their bombs and turned east with the MiG-21s in hot pursuit. The chase went deep into Sinai, until the Soviets disengaged due to diminishing fuel levels, but not before one of them launched an Atoll AAM which exploded close to an A-4, damaging it and forcing its pilot to make and emergency landing at Refidim air base.

The Chel Ha'Avir wanted to give a "lesson" to the Russians and Moderchai Hod obtained authorization from higher levels. An "ambush" of the Soviet interceptors was planned for July 29, although for political considerations, was delayed to the next day.

Four Mirages would simulate a recce flight over the Soviet patrolled area, and as soon as the MiGs scrambled, reinforcements orbiting nearby — which consisted of four additional Mirages and four Phantoms — would join the Mirages. The pilots for this mission were carefully selected from among the best in the Mirage and Phantom squadrons. By 14:14 hours of 30 July 1970, the "recce" Mirages were over the area finishing the first run; about seven minutes later, when starting the second run near the Ras Arab area in the Gulf of Suez, the Soviets appeared. Up to 24 MiG-21s had been scrambled, the majority from Beni Suef, but also from Khatamia and Kon Awshin. The dogfight commenced at about 30,000 ft. First to score was Mirage pilot Asher Snir, who launched both of his Shafrirs, one after the other, against a MiG-21; the sand and brown camouflaged interceptors exploded in a brilliant flash. Miraculously, the pilot ejected and his chute opened almost immediately — he sure had a grand stand view of the air battle. The combat was a little confusing, with some near misses but quickly the highly experienced Israeli pilots got the upper hand the Phantoms dispatched two MiG-21s in rapid succession, being followed by one by a Mirage — the Mirage pilot's name is not known. Then Snir's Mirage was damaged by shrapnel from an Atoll near miss and he had to abort the combat, landing safely some time later at Refidim air base in Sinai. The last kill was for the same unknown Mirage pilot who had achieved his first victory above, splashing his second kill of the day, a Fishbed, with guns. The Soviets disengaged and several minutes later the Mirages and Phantoms returned to their bases, although some, short of fuel, landed at Refidim. But what had happened to the Soviet pilots?

Of the five shot down, two ejected safely, two ejected but were wounded, and one was killed. The Soviets were aggressive and handled their nimble fighters well, but their tactics seemed a little ineffective. However, they little to be disappointed about, as they had fought against a group of aces with a tremendous level of experience. Shortly afterwards a ceasefire came into effect and the July 30 air battle was the last in the Attrition War.

SKIRMISHES

While the Attrition War continued in the Canal Zone, toward the end of June, the situation in the Golan Heights heated up with frequent artillery and tank duels. In fact, this short period of fighting was known as the "three-day War." On June 25 a MiG-21 was claimed, while the next day the score was doubled as two MiG-17s and two MiG-21s were downed, all in air combat.

In fact, the end of the Attrition War did not mean that the war in the air against the Egyptians had stopped, although the first combat was not recorded until two years later. On 13 June 1972, two EAF MiG-21s were downed 40km north of Port Said when they tried to intercept an Israel RF-4E Phantom on a recce sortie, both Egyptian pilots ejected over the sea.

A IAI Nesher, fitted with a 1300 liter drop tank on the centerline, performing a low level barrel roll -quite a dangerous maneuver— pursuing an Egyptian Su-7, note that the Shafrir missile on the right wing rail is missing, used to shot down an Su-7 whose column of smoke can be seen in the background. (IDF)

Egyptian MiG-17 Frescos were common adversaries for Israeli Mirages during the Six Day, to the Attrition and ending in the Yom Kippur Wars, it was a though adversary, but most of the times came second best in air-to-air engagements. (P.Steinemann)

During this period, attentions was on the northern front, especially when the Syrian A.F. tried to defend PLO targets in Lebanon and Syria which were being subjected to Chel Ha'Avir attacks. On 9 September 1972, three SAF Su-7s were downed in air combat after performing a counterattack against IDF positions in the Golan Heights, two were shot down by Mirages and one by Phantoms. On November 9, two Syrian MiG-21s were shot down, one of them by a Mirage with an AAM. On November 21 and in two separate combats, six Syrian MiG-21s were shot down — two and four respectively — five of them by Mirages and one by an F-4E.

IAI Nesher S from 117 Squadron return to its base in Sinai after a successful engagement during the Yom Kippur War, note the empty AAM launch rails and lack of centerline fuel tank. (IDF)

The new year commenced with increasing tension in the Golan Heights, and on 8 January 1973 another air battle developed in which two Syrian MiG-21s were shot down without loss. This was the Nesher baptism of fire and first kills. In fact the pilot which claimed one of the two Fishbeds said afterwards that the MiG-21 was generally superior to the Nesher in its dogfighting qualities.

But the biggest combat on that front was also the last before the Yom Kippur War, taking place on 13 September 1973. Four F/RF-4E Phantoms on a recce sortie, top cover being provided by eight Mirages, were intercepted by 16 Syrian MiG-21s off the Syrian coast, over international waters. Nine MiG-21s were claimed as well as a single Mirage (by an Atoll), its pilot ejected over the Mediterranean and a CH-53 was sent to the rescue. While the Mirage pilot was being winched-up, MiG-21s tried to interfere with the operation and another four were shot out of the sky. The CH-53 also took a Syrian pilot out of the water. Ten of the kills were recorded by Mirage IIICJs, two by IAI Neshers and one by an F-4E Phantom.

According to the figures published by the Israeli Air Force Magazine (which can be considered official), during the Attrition War in 97 air combats, 101 kills were claimed for the loss of five Israeli aircraft. 117 Squadron was credited with 29 kills during those years, while 101 Squadron (which celebrated its 25th Anniversary in May 1973) was credited with a total of 92 kills obtained with Avia S199s, Spitfires, Mustangs, Mystères and Mirages.

THE YOM KIPPUR WAR

The third war for Israeli Mirages started on Saturday, 6 October 1973, when at 14:00 hours on the Jewish Day of Atonement, a simultaneous surprise attack developed in two fronts, the Golan Heights in the north and the Suez Canal in the south. In the latter, the attack commenced with a 1,000-gun barrage against Israeli positions and a crossing of the Canal at four points by the Egyptian infantry, while the EAF attacked Israeli Army targets in Sinai.

The Israeli Air Force had at the time three Mirage/Nesher squadrons: 101 at Hatzor, and 117 at Ramat David — both with mixed equipment (119 Squadron had already converted to the F-4E), and 113 Squadron at Etzion in the Sinai peninsula equipped only with Neshers. Etzion air base was located in the southeast peninsula and west of the Israeli city of Eilat. Other Sinai bases which housed Mirage/Neshers detachments were Refidim (El Arish), Ophir and Eitam, the latter opened in 1977. Eventually all were returned to Egypt in 1982. Returning to 101 and 117 Squadrons, both became a conglomerate of the delta winged fighter, with Mirage IIICJs in their original form, with and without radar, some already reengined with the Atar 9C, some even rebuilt using IAI manufactured components.

About the Nesher, some used French supplied engines (assembled in Israel), others used engines partly manufactured by Beit Shemesh, and the latest batches used all-Israeli Atar, a truly cosmopolitan group. For operational and training sorties, and whenever possible, the different batches flew together, while the fastest and less problem prone Mirage/Neshers were given to the hottest pilots. The main tasks of both fighters were interception and air superiority, being credited with more than half of the 277 confirmed kills claimed by the Chel Ha'Avir including those by AA fire and Hawk SAMS. 102 kills were attributed to the Shafrir II air-to-air missile operated only by the Mirage and Nesher. 117 Squadron was credited with 55 victories, 101 Squadron with 48 and 113 Squadron with 42. In exchange, the Mirage/Nesher squadron sustained ten losses, mostly caused by Arab fighters, six from 101 Squadron, four from 117 Squadron and nil from 113 Squadron.

Nesher S number 15 during turn around in an IDAF air base "somewhere in Sinai" (IDF)

IAI Nesher S serial number 528 from 117 Squadron, note the to "kill" roundels and identification triangles. (author collection)

As there is a problem in exactly differentiating between Mirage and Nesher operations, in the air battle account which follow, when it is said Nesher then it is Nesher, but when it is said Mirage it may be also Nesher. During the length of the conflict an identification problem emerged when Libyan Arab Air Force Mi-

Mirage pilot Asher Snir, the second Israeli top scoring ace, with 13 kills all in Mirages, died in 1986 after a painful illness with the rank of Tat Allof —Lt. General— (IDF)

Mirage IIICJ in flight high over Israel. In 1977 after the arrival of the first McDonnell Douglas F-15 Eagles, the Mirages and Neshers were in charge of pure air superiority missions. (IDF)

One of the top scorers was this Mirage IIICJ with eleven kills. (IDF)

rage 5s operated from Egyptian bases and thus, to avoid mistakes, large yellow/black triangles were painted on the upper and lower surfaces of both wings and both sides of the vertical fin. These were deleted during the late seventies, when the Egyptian Air Force having the same problem with the Libyans due to their border clashes, painted similar yellow-orange identification markings to Egyptian Mirages.

Mirage IIICJ number 107 in maintenance, the groundcrew are removing the Martin-Baker ejection seat. (IDF)

On the first day of the war, the Syrian and Egyptian air forces made a considerable effort, launching numerous sorties, but during the following days mainly left the front battle area to the Air Defence (missiles/guns) arrays to achieve local air superiority. This situation changed when the Israeli armed forces regained the initiative, occupied the Golan Heights again and invaded Syrian territory and crossed the Suez Canal. The hard-pressed Egyptians and Syrians, with their air defence systems attritioned, had to call on their fighters for help. The air activity increased considerably.

The first combat on October 6 took place when a formation of four Neshers from 113 Squadron, patrolling the Tasa-Om-Mashiba area, intercepted eight Egyptian Su-7s at 14:00 hours shooting down four, among them was the late President Sadat's brother, who was killed. Rav Seren Irmi, the Nesher squadron vicecommander, got two, one with guns and the other with an AAM. His wingman bagged a third with guns also and the fourth was downed by the number three Nesher with a Shafrir.

Ten minutes later, Rav Seren (reserve) Eythan Carmi, attached to 101 Squadron, in Mirage number 159 (which up to then was credited with 12 aircraft kills), intercepted an AS 5 Kelt cruise missile launched by an Egyptian Air Force Tu-16 bound for Tel-Aviv and shot it down into the sea with AAMs (which missed) and guns. On the northern front, 117 Squadron was credited on that day with only one Syrian MiG. During the late afternoon/early evening, the Egyptians launched a large heliborne operation to insert commando forces at key points in Sinai with large numbers of Mil Mi-8 Hip helicopters. These helicopters proved extremely difficult targets for intercepting Neshers and only one or two were claimed on that day. A Mirage was lost on the northern front, apparently by a technical malfunction, just before the pilot had downed a Syrian MiG-21. He ejected near the Lebanese coast and was rescued by an Israeli Navy missile boat.

On October 7 the Chel Ha'avir undertook a major but confused effort to attack Arab Air Defence (missile/gun) systems, first in the south and then in the north. During the morning 101 Squadron's vicecommander was lightly wounded in a take-off accident in Mirage number 159 which was badly damaged, but was rebuilt afterwards by IAI. This squadron lost two of their Mi-

Mirage IIICJs from 190 Squadron (below) in flight and during an engine run after a maintenance check (above). (IDF)

rages in air combat that day, both in the southern front and after their pilots achieved kills. One of the pilots was Eytan Carmi, who previously had downed two EAF aircraft over Port Said. The two jocks ejected safely, landing in Sinai and being recovered by helicopters. 117 Squadron claimed eleven Syrian Su-7s and Migs, but 113 Squadron was not credited with even a single kill and its commanding officer urged the Chel Ha'Avir Staff to assign his squadron ground attack missions.

The first ground attack mission took place in the early morning of the next day, October 8, when Neshers attacked the Jabal-Druz area in Syria, but were sent immediately afterwards to air-to-air missions in the Suez Canal area. There the Neshers engaged in at least two air-to-air combats, the first in the morning in which the squadron commander downed one MiG-17 and his wingman two of the same type. The other combat took place in the afternoon over the Nile Delta, again the C.O. scored, this time a MiG-21. Asaf Ben-Nun (an IAI test pilot flying as a reservist) got a second Fishbed, while a Mirage from another squadron was lost in this engagement. That morning, at 09:00 hours, a pair of Neshers detached on interception alert at Refidim air base (ex-Bir Gafgafa) in central Sinai. They scrambled to intercept six low flying Egyptian Su-7s heading in to attack the base. Meeting the fast Sukhois, the Neshers closed-in. At about the 1,000-meter range, the Israeli leader launched a Shafrir, which failed as the Egyptian saw it and made a panic break. Not to be left without a possible kill, the Nesher pilot also turned hard and again positioned the fighter in his sights, at a range of about 600 meters. He launched his second Shafrir. This time the missile found its target and the rear part of the Su-7 went up in a brilliant flash, seconds later crashing on the desert floor. Meanwhile, his wingman had bagged other Su-7 with a Shafrir, and a third fighter crashed into the ground while taking evasive action, without any ordnance fired into him.

During the late afternoon, Sgan Allof Oded Morom — an ex-Mirage sqn. C.O. — intercepted four Iraqi Air Force Hunters (two squadrons of which were temporarily deployed to Egypt) attacking Tasa Camps. Morom shot down one but some flying debris fragmented his canopy and he had to make an emergency landing back at his base. Another Mirage from the same formation downed two Hunters, while four were claimed by AA fire. The Iraqis admitted the loss of all four Hunters on that sortie. Other kills credited to the same squadron on that day were one MiG-17, four Su-7s and three MiG-21s. Meanwhile 117 Squadron had a weak journey with no kills to its credit. Also on October 7 and 8, Mirages and Neshers flew several ground support missions, strafing Egyptian bridges on the Suez Canal.

On October 9, Mirages escorted 16 F-4E Phantoms to attack the Syrian Army and Air Force Headquarters buildings in Damascus. In fact, during that stage of the war the main attention was on the northern front. On this day, 117 Squadron scored six victories and eight on the next day. On October 13, Sgan Allof Avi Lanir, 101 Squadron's C.O., ejected after an air combat over Syria, the unfortunate pilot was captured and tortured to death by the Syrians. He was replaced by his vicecommander, who had been wounded in a take-off accident on October 7. This pilot insisted that in order to lead the squadron he should fly and fight. With only one hand in working order, he shot down a Syrian MiG shortly afterwards. On October 11 Colonel Mamdoch El-Malt, an EAF MiG-21 pilot, claimed to have shot down a Mirage on that date near Fayid. Later on during the same combat, he was downed and taken prisoner. The next day, Seren Ami Rockach shot down two Syrian MiG-21s while flying a Mirage. On his way home he was hit by AA fire (also there is the possibility that his fighter was badly damaged during the dogfight) and ejected. While trying to escape from Syrian soldiers he was shot in the chest, being later taken prisoner seriously injured. He was released by the Syrians on 1 June 1974.

On October 14 four Neshers from 113 Squadron patrolling over the Port Said area were sent to the Nile Delta to help several F-4Es which had been bounced by MiGs. The squadron C.O. shot down two, one with a Shafrir and the other with guns. Asaf Ben-Nun got one with guns and a third Nesher destroyed another MiG-21 with a Shafrir missile. During the afternoon of the next day, four Neshers escorted four Phantoms to attack Damascus International Airport, after a previous attack had been intercepted over the area. The sortie was uneventful except for a slightly damaged Nesher from an SA 3 SAM exploding too close for comfort.

Toward the last phase of the conflict, large air battles became more and more common in the southern front as the Israelis had, in a brilliant and daring military operation, established a bridgehead on the mainland side of the Egyptian territory. Sgan Allof (Reserve) Giora Epstein was at the time a Staff Officer flying on emergency appointment with one of the Mirage squadrons, and it was during this later stage of the war that he usually was in the right place at the right time, as his score rocketed skyward. He shot down with guns a Mi-8 helicopter attacking the bridgehead established by the Israeli Army on the West Bank of the Suez Canal in twilight with napalm. On October 19 in about the same area and in two separate sorties he was credited with two Su-7s and two Su-20s (see his narrative in the last part of this chapter). On October 20, again around the same area, he shot down four MiGs (three with guns and one with an AAM) in a single sortie during the afternoon.

Two days previously, a curious incident took place when a pair of Mirages were scrambled from Refidim air base to intercept attacking Egyptian Air Force MiG-17s. One of the pilots shot down two MiG-17s, both with Shafrir and Sidewinder respectively (during some missions the Mirages and Neshers carried one missile of each type).

The second Mirage pilot flew directly into the fireball of one of the Egyptians and emerged black overall, its pilot seeing almost nothing. The pair aborted the combat and returned safely to Refidim air base.

On October 21 four Neshers from 113 Squadron patrolled over the Faid area and intercepted Libyan Mirage 5s. At least one was shot down with a Sidewinder, which damaged it, and was finished with the guns. Previously in a battle over the Bordavil area, two 101 Squadron Mirages intercepted a flight of four Libyan Mirage 5s which were over the sea and heading into Sinai. When jumped, they broke hard to the right, jettisoning their underwing loads (bombs and tanks) into the sea; the number four man overmaneuvered and hit the water with the right wingtip, the crash raising a great plume of water. After a chase to the west, the Israeli wingman shot down two in quick succession, one with guns and the second with a Shafrir. Both crashed into the sea, the only survivor made good his escape at very low levels.

The afternoon of October 24 saw the last large air combat during the war, in which Giora Epstein achieved three more kills, two with AAMs and one with guns. Epstein raised his personal score to 17, twelve of which were recorded during the Yom Kippur War, emerging as the Chel Ha'Avir leading ace, and possibly the world's top jet ace. Shortly afterwards a ceasefire came into effect.

The Mirage/Nesher squadrons had been mainly committed to the air-to-air role, suffering much less than the F-4, A-4 and Super Mystère units and, in consequence, the Air Defence arrays were much less traumatic to them. The last Mirage loss in combat in Chel Ha'Avir service took place late in October, just after the ceasefire, when Eytan Carmi of 101 Squadron went over Syria for a recce sortie. Hit by a SAM, his Mirage was badly damaged but managed to head west. Carmi ejected over the Mediterranean Sea near Akko Beach, being rescued by an Israeli Navy patrol boat. This was the second time in less than a month that he had to abandon his fighter in flight!

Mirage IIIBJ in 190 Squadron markings, note the supersonic 600-liter drop tank fitted on the centerline. (IDF)

Mirage IIICJ number 111 is the one assigned to the IDFAF Museum at Hatzerim air base in the Negev, the 13 kills were painted for effect as this particular aircraft did not achieve such a score.

After the Yom Kippur War and with the introduction of the Kfir in 1975 (101 Squadron), the McDonnell Douglas F-15 Eagle late in 1976 (with the newly activated 133 Squadron at Tel Nof) and the General Dynamics F-16 Fighting Falcon in 1970 (117 Squadron), the Mirage and Neshers were relegated to second place among Chel Ha'Avir interceptors. Between 1978 and 1981, 35 Neshers Ss and four Nesher Ts were phased out and sold to Argentina, while curiously, the Mirage IIIs, which were older, served longer with 190 Squadron, which had received them in 1978. The last conflict in which the Mirage IIIs were involved was during the Lebanon War in June 1982, being used only for CAP missions over northern Israel and in strip alert readiness, without achieving any results. Their pilots watched in awe as their F-15 and F-16 buddies badly mauled the Syrian Air Force. Shortly afterwards the Mirages were withdrawn from service and sold to the Fuerza Aérea Argentina, while 190 Squadron reequipped with Kfirs and, several years later, with F-16 Fighting Falcons.

YOM KIPPUR WAR PILOT NARRATIVES

Allof Mishne, a retired Chel Ha'Avir fighter pilot, tells here about a hard fought dogfight with some Syrian Su-7s over the Golan Heights. His first kill took place on 6 June 1967, during the Six-Day War. It was an Egyptian MiG-19, downed while flying a 101 Squadron Mirage.

". . . During a CAP over the Golan Heights my wingman spotted two Syrian Su-7B strike fighters coming in at twelve o'clock high. A quick warning shout brought the two Mirages turning to meet them. The Sukhois barrelled downwards, sweeping low over the Golan Heights, followed closely by puffs of smoke as the rockets exploded below them on the ground. Pulling-up sharply, the two intruders groped desperately for altitude. We hit the afterburners and rolled into position behind the leading Sukhoi, just as it started a strafing run. As the Syrian pilot saw the Mirages coming up behind him, he pulled up sharply, breaking hard to starboard, then turned to meet us head-on with his 30mm cannon. I tried to get inside the Syrian's turn. We were now to hill top level. The Su-7 vanished momentarily as we rounded a small hill, we were turning very tight. Brown earth and blue skies mingled as we followed the low level chase. My Mirage closed in on the Sukhoi's tail, but the "g" forces were too high. The Syrian was either an ace or a madman. By now the fight was down to zero altitude, twisting and turning in the wadis round Mt. Hermon. The Syrian pulled some crazy tricks, throwing his aircraft into sharp angles. He was a good pilot. I was sweating hard trying to keep up with my prey. Suddenly I got my chance. Only for a moment, but I managed to get a quick burst. The Syrian broke hard, a small fire blossoming bright red below the wingroot. Still he flew on. By now the fight was on again over the frontline, with the Syrian frantically attempting to shake me off and dash home. I saw the fire in the Su-7 spread. I drew closer watching the pilot struggling in his cockpit. Suddenly it blew up, engulfing my Mirage in a cloud of smoke. I was blinded, and felt the elevons jerk the stick inside my hand. It felt like hitting a brick wall in a racing car. The aircraft kicked back like a mule, as I tried to roll and turn at low speed. Struggling to keep my fighter on an even keel, I managed to get clear of the fireball and flying debris. The fight was over . . ."

Sgan Allof Giora Epstein was the Chel Ha'Avir top ace and the Yom Kippur War's top scorer. In the following lines he describes some of his air to air kills.

". . . We flew on a sortie to the northern Bitter Lake. The Egyptians were attacking in masses. They would arrive at low level, in large waves of aircraft, climb while turning to attack parallel to the Canal strip, then head for home quickly. We had little time to catch them. As we reached the Israeli bridgehead near Deversoirs, we watched a flight of five Su-7s pulling hard left in order to pounce upon our forces below. We came from behind. I launched a missile on the last man, which exploded immediately. The leading Sukhoi pair dropped their bombs. The leader of the second pair did not make it and turned for home with me in hot pursuit. I chased him at low level, my aircraft shaking from the speed, making it difficult to keep him in the sight. Just before entering the SAM zone I managed to hit him.

Back at Refidim, we were soon scrambled again, and I shot down two Su-20s over the Canal. During this later engagement a curious thing happened to me. Apart from the Mirages which had taken part in the battle, we had some Phantoms in the area. A four-ship F-4E formation got in behind the first Sukhois in order to make a pass, sandwiched in between the first and second enemy formations diving down. One Sukhoi threatened a Phantom just in front of him. My wingman jockeyed into position, but we were afraid that the Egyptian would hit the Phantom before we could stop him. I tried to warn the F-4E driver, but did not know his call-sign. As I got excited, I yelled "Phantom, break!!" The man did . . . instantly, and my wingman shot the Sukhoi down. I returned to Refidim. Morale was high, we had downed eleven aircraft that morning.

The next day we were directed to Jebel Oubeid near Suez. At 10,000 ft. we detected a pair of MiG-21s heading north in a wide turn. We dropped our tanks and gave chase. I maneuvered into position at the 1,000 meter range and launched a Shafrir missile. The MiG exploded, but now I realized I had fallen into a trap. Still occupied, I saw, out of the corner of my eye, scores of MiGs climbing hard to meet us. They looked like mushrooms in the sky. We parted into two pairs and dived to engage. The melee started almost immediately. My wingman closed onto one MiG, sending a missile after him, but due to mechanical problems had to break the engagement. Two other Mirages were forced to disengage due to low fuel. One more MiG was downed before they left.

I was still chasing my MiG all that time, and that man was good. He did all sorts of crazy things with his airplane at low level, split-s'sing so low that he was throwing sand every time he pulled out from his dives. I was right behind him, glued to his tail. Meanwhile some other MiGs tried to get behind to shoot me off his tail. But if I could not manage to bring my sights on the man in front, what chance could they have against me? At least, when the MiG pulled up, I got my break and managed to put some rounds into him. He dived into the ground. As I pulled round I saw, just in time, that two MiGs had launched missiles toward me. They missed and passed me close by. I broke sharply and watched as two MiGs launched missiles straight ahead at me! Here I had no time to react. I ducked in my seat instinctively as the missiles passed over me. Luck continued with me. Instantly I turned and found still another pair closing in behind. Things were getting hectic, and I broke away sharply, making for home . . ."

5

Born in Battle, From Nesher to Kfir

One of the consequences of the Six Day War was the French embargo, first on the 50 Mirage 5Js and two 5DJ fighter bombers and from December 1968 on all military equipment to Israel. At the same time, the Israeli industry started an effort to fill the gap, and one of the early products was the Rafael Mahat weapons delivery computer, used to enhance the accuracy of the Mirage in air-to-ground operations. Results were obtained in a test program that involved fifty bombs dropped from different heights and angles. The development phase ended in 1970, but at that time — with the arrival of the F-4E Phantom — the Mirage role was changed again to air superiority and the Mahat was not installed.

Another product was the Elta EL/M-2001 ranging radar designed specially for air-to-air combat at low altitude, entering service with the Chel Ha'Avir only in early 1973, being fitted to some of the Neshers then in service.

The most interesting outcome of the embargo was the Nesher (Eagle). Initial trials were made in a Mirage IIIBJ which acted as a test vehicle with monitoring equipment installed in the rear cockpit. This was, in those years, quite an ambitious program for such a small country as Israel. At one time or another, the Israeli technicians, which had collaborated with Dassault in the development of the Mirage 5 and had returned to Israel, received the unofficial help from Armée de l'Air and Dassault engineers, producing exact manufacturing drawings and the necessary tools and jigs. Also, the Mossad "contracted" the services of a Swiss engineer, Alfred Frauenknecht, who worked for the Sulzer company, then producing the Atar 9C engine under license. The Sulzer employee supplied the Israelis with loads of Mirage and Atar 9C blueprints. However, he was caught by the Swiss police and sentenced to four years in prison.

The first IDFAF unit to equip with the IAI Kfir was 101 Squadron, which in mid-1975 discarded its Mirages and Neshers to other units and reequipped with the first variant, the C1, although two years later converted to the much improved C2. (IAI)

The Kfir was designed as a ground attack fighter, thus its weapons load capability and diversity was augmented to that of the Nesher as can be clearly appreciated in this picture of a C2 showing its armory. (IAI)

The first prototype of the Nesher made its initial flight in September 1969 with IAI Chief Test Pilot Dani Shapira at the controls, while the first series production example made its maiden flight on 8 January 1971 with the same pilot. The Nesher was powered by a copy of the Atar 9C3 which had been manufactured since 1968 by Bet Shemesh (a subsidiary of IAI). The fighter was at first delivered with only basic avionics. It was almost an exact duplicate of the French Mirage 5, with the electronics bay repositioned in the nose, and its original place occupied by a 500-litre fuel tank called the Forward Upper Fuel Tank Cell. It also had a Martin Baker JM6 zero-zero ejection seat. The electronic equipment fit lagged behind during the development phase and the aircraft were retrofitted as this equipment became available. The Chel Ha'Avir received 51 Nesher Ss. Deliveries started during 1971 and about 40 were in service on the eve of the Yom Kippur War in October 1973. The first recipients were the Mirage III units, 101 and 117 Squadrons, which operated both types (119 Squadron had converted to F-4Es in 1971). In late 1972, 113 Squadron discarded its obsolete Dassault Ouragans and converted to Neshers based at Etzion AB in Sinai. The prime role of these units during the October 1973 war was interception and air superiority. 113 Sqn. carried a black/yellow checkered rudder in some of its aircraft as well as the "Wasp" badge on the fin. Other units to receive the Nesher were 111 and 190 Squadrons. Ten two-seat conversion trainers, named Nesher Ts were built by IAI, while the last of the Mirage 5 copies was delivered by IAI in 1975, just at the start of Kfir production.

The international aviation press first became aware of an Israeli produced combat aircraft just before the October 1973 War; but, the Israelis guarded one of their best kept secrets (although the Sulzer employee scandal in Switzerland made the matters more difficult). The obsessed, security conscious Israeli military fabricated a decoy story about a project called "Barak" (Lightning) which was an Israeli-produced Mirage III with a J79 engine and which was already in Chel Ha'Avir service. The authentic fact is that before the Yom Kippur War, Nesher number 712 was converted as the first Kfir, making its maiden flight in June 1973 as is narrated later.

The Neshers spent relatively few years in front line service. With Kfir production gaining momentum, the air force staff considered that Neshers were no longer needed, so in 1978/1979 24 single-seaters and two-seaters were sold and delivered to Argentina as Dagger As and Bs. These were followed by a second batch in 1981 of 11 and 2 respectively. The remaining were w/o in accidents, combat or converted to Kfir, while at least one is preserved in the Chel Ha'Avir Museum at Hatzerim AB in the Negev.

While all this took place, the availability of the Israeli produced Atar 9C engine made matters easier to start a Mirage III mid-life-update program (MLU). This began during the late sixties and continued even after the end of the Yom Kippur War. The MLU was applied to all the aircraft which passed through the periodic depot maintenance schedule. Thus the Atar 9B3 was replaced by the more reliable 9C3, the electronic fit was upgraded, including radar warning receivers (RWR), capability to carry four AIM-9s or Shafrir II in place of two, new wing skin and spars, while the troublesome Cyrano Ibis fire control radar was removed from most of the aircraft (the ones which kept it had it jammed to act merely as a rangefinder, others were fitted with Elta ranging only radars). During the late seventies some of the Mirages — which up to then had been painted in a desert-type lizard scheme, replacing the silver finish after the end of the Six Day War — received a new gray overall air superiority scheme. Progressively, the Mirage started to disappear from the first line inventory of the Chel Ha'Avir, first was 101 Squadron which converted to Kfirs in 1975, followed by 113 which also received Kfirs in 1977, 117 Squadron which got F-16A/Bs in 1979 and finally 190 Squadron which also got Fighting Falcons during the last months of 1982. The majority of survivors did not remain in Israel as 19 Mirage IIICJs and three IIIBJs were supplied to Argentina in December 1982.

KFIR, ISRAEL'S LION CUB

The change in the Chel Ha'Avir from French to American equipment in the late 1960s was related both to political and operational reasoning. In less than five years, the Air Force suffered a dramatic change. From an all-French front line equipped air arm, its front line inventory became almost totally dependent on Ameri-

Production line of the Kfir at the IAI assembly plant followed that of the Nesher in 1975. (IAI)

Kfir number 714 was originally built as a C1, but later in 1975 was modified to become the full scale development article of the follow-up variant, the Kfir C2, its main different was its new canard winglets. (IAI)

can warplanes. The U.S. aircraft were especially appreciated for their efficient propulsion systems and avionics but they lacked in aerodynamic refinements. An example is the Phantom, which relied on sheer power. The opposite was the case concerning French warplanes. Their propulsion systems were provided less thrust than the American equivalents, while most of their avionics were completely unreliable, but concerning aerodynamics, the Mirage was a real masterpiece. Hence, the ingenious Israelis had the temptation to merge the best of both worlds, American engine and avionics with French airframe.

That idea was first raised in 1968, even before the first F-4Es arrived in Israel but some time after the A-4Hs entered service with two Israeli squadrons (109 and 110). Thus, with the supply of the Phantom's General Electric J79 still not assured, IAI engineers evaluated the Rolls-Royce Spey and the P&W TF30 turbofans as well as the J79. The Spey was discarded quite early on, since its performance was not adequate and its availability was not assured. The dimensions and weight of the TF30 were such that its mismatch also became obvious very soon. It then became apparent that the F-4E would indeed enter Chel Ha'Avir service and this, coupled with the general shift toward U.S. equipment and the roughly similar diameter of the J79 and the Atar 9C, resulted in the adoption of the American turbojet as the power source of a new project which had a number of codenames.

"Salvo" (Courtain) was a codename used by IAI in its correspondence with GE, and that codename was leaked to the press. Another one provided to the press as an act of disinformation as "Barak" which curiously was close to the actual codename used by IAI for all its family of Mirage 5 derivatives, which became Project Ra'am (Thunder). Yet one more codename used after-

A line-up of 101 Squadron Kfir C2s without the classic markings of the unit, the red/white stripped rudder and the winged skull insignia.(IAI)

With full afterburning power from its J79 engine Kfir C2 takes off from Ramat David air base loaded with three auxiliary fuel tanks. The C2 was a good ground-attack fighter and paved the way for the "digital" C7. (IDF)

wards by the Chel Ha'Avir for its enhanced requirement was Rakdan (Dancer). In the end, none were used for the operational versions.

At first the idea was to fit an updated engine in the French fighter to increase its performance. No equipment changes were proposed, the fighter was simply to be an updated version of the Nesher. This had sounded simple but, in fact, a total change of powerplant type was and is a demanding task. Although the diameter of the two engines was similar, all other parameters were totally different. Length, weight distribution, attachment points, mass air flow, temperature range and control systems all had to be changed. As a consequence, a major redesign involving four main aspects of the aircraft was required.

The physical installation of the J79 instead of the Atar resulted in redesign of the structural/powerplant attachment points, the strengthening of those areas, and the internal enlargement of the fuselage without any external changes. Secondly the air inlets and the movement laws of the variable geometry shock cones had to be completely redesigned. Thirdly, the difference in operating temperatures led to the installation of thermal blankets around the powerplant within the limited space available, and the replacement of certain aluminum components for titanium parts in areas prone to very high temperatures. Finally, all the accessories (fuel system, hydraulic system, air conditioning system and generators) had to be redesigned completely or replaced.

Other externally evident changes involved the shape of the afterbody and the addition of small air inlets for cooling air. In short, the project did not only involve the simple action of engine replacement but a major redesign of the aircraft. To test the new engine/airframe combination, it was decided to install a J79 turbojet in a Mirage III airframe. A Mirage IIIBJ was selected because the extra space available in the rear cockpit could be used for test instrumentation, as already had been demonstrated with the Nesher project. Another reason was that the assignment of a

The first Kfir C2, number 714, in its element, showing clearly its modifications, like the canard foreplanes and wing's dogtooth leading edges, both items greatly improved maneuverability and low speed controllability. (IAI)

The Kfir TC2 was a combat trainer development of the basic C2, fuselage length was slightly increased, while the second cockpit forced the removal of the upper fuel cell. First flight took place late in 1980 and a small production run followed. (IAI)

two seater affected the first line strength of the air force less. In April 1969, Mirage IIIBJ (9)88, was loaned to IAI, never again to return to Chel Ha'Avir service. It became a test bed and demonstrator, nicknamed the Technolog, before it finally retired to the Museum at Hatzerim, where it remains today. The J79-GE-17 to be installed in the Technolog was taken from spares obtained in the first F-4E deal and thus had to be modified to suit single engine operation. Later J79s for production aircraft were procured specifically for that purpose and assembled by IAI, which introduced several refinements including improved combustion chambers to lower the smoke signature and a titanium heat shield to cover the afterburner.

Test pilot Dani Shapira was at the controls of the Technolog during the maiden flight on 21 September 1970. Shapira was chased by Asaf Ben-Nun, the IAI project test pilot, and it was Shapira who asked Ben-Nun to check his tail for signs of fire as the temperature warning light in his cockpit came on. Ben-Nun reported that he was unable to see any fire and Shapira concluded correctly that the fault was with the warning system. Once that problem was rectified and the flight envelope was thoroughly explored, the Technolog proved its superiority in mock air combats against Mirages and F-4Es, the latter already in service with two squadrons (107 and 201). Being a cleaned-up and highly boosted Mirage, it is not strange that the pilots who flew the Technolog left its cockpit amazed at its outstanding dogfighting performance. The fact that the series production version was not so agile came as a surprise to many who were used to operating fighters in the air-to-air role, rather than as mud-movers.

This was mainly attributed to changes in operational requirements of the Chel Ha'Avir. The need for an air superiority fighter was by then not so urgent, as A-4s and F-4s were being supplied in quantity, and the first Neshers were about to leave the IAI production line. Thus, the target delivery for the new aircraft, November 1973, could be relaxed. At the same time, the changing threat and the lessons of the Attrition War also contributed to that shift as the AA defenses became much more dangerous and higher emphasis had to be placed on survivability.

The new generation of combat aircraft, namely the U.S. F-14 and F-15 air superiority fighters, made it obvious that the new Israeli fighter was based on technology which dated back to the 1950s. So the emphasis sifted again to a multirole fighter. To enhance survivability over a future battlefield, a major redesign had to be made. This included the relocation of systems, a change of avionics, introduction of a weapons delivery system, addition of a radar and the option to carry higher payloads. It was a shift from the simple French avionics and systems to the considerably more advanced technologies of U.S. and Israeli origin. This led to a decision in November 1971 to continue with the construction of what was basically a J79 engined Nesher, termed by IAI the Ra'am A, while working on the definitive version which was at first codenamed by the Air Force as Project Rakdan and then renamed Kfir (Lion Cub).

The Ra'am A, number 788, was a modified Nesher that made its maiden flight on 4 June 1973 with project test pilot Asaf Ben-Nun at the controls. The progress of flight testing of this prototype was impressive and by 21 June 1973 the first supersonic flight was accomplished. A speed of Mach 2.4 was achieved on December 10. Ra'am A was used to test the combination of the American powerplant and the Israel improved French designed airframe. It differed from the Technolog in the design of the auxiliary cooling air inlets attached to the rear fuselage. Four small inlets surrounded the fuselage while a single larger inlet was installed in the lengthened base of the fin.

The next prototype was yet another modified Nesher, 712, which first flew on August 7, 1974. It differed from the definitive production Kfir mainly in the shape of the nose, due to the fact that the Elta EL/M-2001B rangefinding radar was not installed. The Elta radar, which entered production during 1973, represented a significant improvement over previous systems and its ability to operate at low altitude, in spite of heavy ground clutter, was equally important. While flight testing was proceeding, voices from within the Chel Ha'Avir were heard expressing doubts about the project. The redesign and shift of emphasis resulted in a higher gross weight which degraded air-to-air performance in comparison with the Nesher, which itself was inferior to the original Mirage III. It also resulted in moving forward the center of gravity (CG) and,

With its serial number and unit insignia taped over, a Kfir C7 from 101 Squadron taxies to its dispersal at Hatzerim air base after a flight demonstration. (IDF)

Kfir C7 from 111 Squadron in the later all-gray scheme, being loaded with a diverse weapons load, rockets, bombs and Sidewinders. (IAI)

since its Aerodynamic center (AC) remained in its original position, the aircraft became more stable and less responsive. Another problem was that fuel was consumed in higher rates than ever before. Although the specific fuel consumption of the J79 was lower than that of the Atar 9C, drag on the new aircraft was higher and more thrust was required to overcome it.

By July 1974 the Chel Ha'Avir C.O., Allof Beni Peled, clearly stated that the Air Force was evaluating two types of fighters: an air superiority fighter, either the F-14 or F-15, and a new lightweight multirole fighter with the YF-16 and YF-17 in the running. It was quite plain that the supply of such advanced combat aircraft was assured, although the timing had to be negotiated. It therefore seemed that the Kfir lost its support within the Chel Ha'Avir although the program continued mainly as a result of the support of the Minister of Defence, Shimon Peres.

On 14 April 1975, the first series production Kfir, number 724, was handed over to the Air Force by IAI. During the ceremony the prototype made a short display flight but 724 was not flown. Since production of the Nesher had just ended, the first production Kfir simply followed on a lightly modified production line. Although the transition from the production of Neshers to Kfirs was smooth and mostly troublefree, this was certainly not the case as far as entry into service was concerned. The first Chel Ha'Avir unit to accept the new Israeli fighter was again 101 Squadron, under the command of Sgan Allof Hertzel Bodinger. The squadron relinquished its remaining Mirages and Neshers to other units and began a working-up period with the Kfirs. This, however, proved to be somewhat problematical as the new aircraft soon acquired an unjustified reputation as a failure. In the air-to-air role, the Kfir C1 was indeed a loser while even in the air-to-ground role its performance was poor. Pilots of 101 Squadron, previously an air-superiority unit, disliked their new mount from the outset, although in many ways injustice was done with this fighter since it was tailored to an Air Force specification which placed higher priority on mud-moving operations than on air combat performance. The Kfir C1 was just not designed for air combat, a point that the pilots of what was previously the Israeli premier fighter outfit failed to appreciate.

But Kfir C1s were rolling off the production line and as 101 Squadron completed the transition, a second unit, 113 Squad-

Kfir C7 number 534 loaded with four AIM-9G Sidewinders flying over the tormented landscape of southern Israel. Kfirs will be maintained in service in the IDFAF until the year 2000, when the last squadron is expected to disband or convert to other type. (IAI)

Patrol over the Red Sea, a pair of Kfir C7s in their desert lizard scheme on CAP near the southern Israeli city of Eilat, the waterway to the Red Sea. (IAI)

ron, reequipped during 1976. A third, 109 Squadron, which previously operated A-4H Skyhawks from Ramat David was to operate C1s, being declared operational on 28 July 1977. But the aircraft delivered to the latter squadron were second hand examples obtained from 101 Squadron which, meanwhile, had converted to the improved variant, the Kfir C2. In fact, the aerodynamic deficiencies of the C1 had not been ignored by IAI and, quite early in the program, an effort to reduce drag and improve lift was launched. This resulted in the decision to test the installation of a small canard forward and above the wing behind the air intakes. This was to add lift and move the AC forward to narrow the stability margin, thus improving stability. Secondary benefits were also quite remarkable. The vortices procured by the canard smoothed the air flow over the delta wing and thus enhanced lift and reduced drag, allowing safe operation at higher angles of attack (AOA) and improving maneuverability. The strengthening of airframe section number ten led to the introduction of two mor hard points, to bring the total to nine, quite a respectable numbe even today.

In addition to the canards, two more aerodynamic refinement were introduced. Small strakes on both sides of the nose sectio improved the airflow characteristics while the introduction of saw tooth on the outer section of the wing's leading edges adde to the effect of the canard-induced vortices. The combination c all three aerodynamic improvements enhanced maneuverability especially at the lower end of the speed range, where the Kfir ca be flown under control at 100 knots. A great improvement wa also achieved in the sustained turn rates from 7.8 degrees pe second up to 8.7 and from 16.5 degrees per second up to 18. respectively, all at 15,000 ft. Two more benefits were improve handling and a lower gust response, which is especially impor tant in low level flight — a part of almost every mission profile.

All the aerodynamic refinements were tested on th Technolog, which was first flown in its new guise by Asaf Ben Nun on 16 July 1974. To accelerate the program, the Ra'am was also modified and flew for the first time, with canards an other aerodynamic improvements including extended outboar leading edges on the wings giving a pronounced saw-tooth, o 23 March 1975 with Menahem Shmul at the controls. Barely tw months later, on 25 May, Ra'am A was lost when its pilot had t eject over the Mediterranean but by then the worth of the im provements was proved and Kfir C1 number 714 was also modi fied, making its maiden flight before the end of that year. The nev variant, the Kfir C2, was first shown at the Chel Ha'Avir Day or 20 July 1976 when Dani Shapira demonstrated its improved ma neuverability at Hatzerim air base in the Negev. Deliveries of th Kfir C2 to the Chel Ha'Avir commenced in May 1977, after som 40 C1s had been delivered. The high delivery rate including th reequipment of several more squadrons, including 101, 109, 113 111, 121 and 144, until 1980. Remaining C1s were fitted with small canard plus strakes on both sides of the extreme nose This installation was of limited success and retired C1s were ini tially stored as combat reserve. They had nevertheless achievec

Seen at the IDFAF Open Day in June 1988 is Kfir C7 number 557 showing part of its weaponry, including Python III AAMs, and Shafrir IV, in foreground, blue fuselage. (Aloni)

Kfir C7 number 534 in air-superiority configuration with four AIM-9C Sidewinders and one 600-liter drop tank, aircraft possibly belongs to 144 Squadron. (IAI)

'hree Kfir C7s, this time armed by the Israeli developed and built RAFAEL Python III, infrared guided air-to-air missiles, flying over the Sea of Galilee (also ;nown as Lake Kinneret) on northern Israel. (IAI)

ame as the first Kfirs to perform an operational mission when our C1s from 113 Squadron attacked PLO positions at Tel Aziyah, .ebanon on 9 November 1977. Just over a fortnight later, on November 21, four Kfir C1s of the same squadron were engaged n a totally different mission when they escorted the Egyptian Boeing 707 of President Anwar el Sadat back to Egypt after his istoric visit to Israel.

Until the late 1970s, Kfir squadrons relied on the Mirage and Nesher units as a source of new air crews and, hence, no two-seat conversion trainer was needed. However, with the retirement of the Neshers, which began in 1978 when a first batch of '6 was exported to Argentina, and the anticipated phase out from se of the Mirage in late 1982 when the remaining 22 were also exported to Argentina, an operational requirement for a two seat Kfir was formulated. IAI responded with the Kfir TC2. The basic C2 fuselage was slightly lengthened while the fuel tank immediately behind the pilot was removed to clear space for the second cockpit. The air conditioning system, originally installed below that uel tank, was relocated in the lengthened nose. Other minor changes involved the landing gear, the cockpit canopy and the addition of strakes on the lower sides of the front fuselage ahead of the air intakes.

The first Kfir TC2 made its maiden flight late in 1980 and a small production run followed. By this time the operational use of he Kfir concentrated mainly on attack missions, although it was o achieve its sole air-to-air kill to date on 27 June 1979. On that late a pair of C2s from 101 Squadron joined six F-15s from 148 Squadron in a well planned ambush set up to catch Syrian Air Force interceptors that tried to interfere with the Chel Ha'Avir routine attack activity in Lebanon. The Syrians claimed they launched three formations to intercept the attacking Israeli aircraft, although Chel Ha'Avir pilots observed only eight MiG-21s. According to the Syrian version, their fighters were vectored to intercept the Israeli aircraft when Syrian communications were jammed and their ground control radar disturbed by electronic warfare means.

Attack aircraft turned west from where the CAP formations suddenly climbed from low level to catch the chasing Syrian MiGs. Five MiG-21s were claimed in that combat, of which four were credited to the F-15 pilots but the fifth was credited to one of the 101 Squadron pilots who achieved his kill with a RAFAEL Shafrir 2 AAM.

This particular C2, number 874, is currently on display at the Air Force Museum (see photo). Already combat proven in both air-to-air and air-to-ground warfare, Kfirs were relegated more to the later mission following the arrival of F-16s and additional F-15s from 1980 onwards. During the Lebanon War of 1982 Kfir squadrons mainly flew attack missions. That decision resulted in a third generation Kfir, the main changes being internal and involving an avionics suite which was a great improvement over the previous variants.

On 31 July 1983, the first Kfir C7 was handed over to the Chel Ha'Avir and entered service with 111 Squadron which became the first unit to operate the type. The ground attack capability of the new variant was vastly improved, mainly due to the new avionics suite but also by the replacement of the J79-GE-17 by the J79-IAI-J1E with a combat plus system that increased take-off thrust from 17,750 lb. to 18,750 lb. A strengthened landing gear was introduced to deal with the higher gross weight that became possible due to the increase in thrust.

The main reason for the higher effectiveness of the C7 in the attack mission was the replacement of the avionics and the rearrangement of the cockpit layout to suit the HOTAS (Hands On Throttle and Stick) concept. The heart of the new avionics is the IAI/Elbit WDNS 341 which includes a Tamam central air data computer and Inertial Measurement Unit (IMU), an Elbit mission computer and Stores Management System (SMS), an El-Op HUD and an Elta EL/M-2001B ranging radar. Other contributions to the combat capability of the new variant are an Elisra RWR, two Elta transceivers and a VCR.

Entry into service of the Kfir C7 began some time after the decision had been made to produce the IAI Lavi multirole combat aircraft as its successor. It was originally planned that production of the Lavi would follow directly on from Kfir, much in the same way as the Kfir replaced the Nesher at the Lahav production line. It soon became evident that this was not a realistic possibility and that a gap in the production line was inevitable. Although Kfir production was slowed down, the production line was formally closed in the spring of 1986 after 212 Kfirs of all four variants, C1, C2, TC2 and C7, were produced.

Kfir C7 cockpit. The last Kfir customer is the Sri Lanka air force (SLAF), which early in 1996 received four C2s and two TC2s from IAI taken from IDF/AF surplus stocks and upgraded, being issued to No. 10 Fighter Squadron at Katunayake AB, west of the island, which put them into immediate operational use in CAS missions against the Tamil Tigers guerrillas in the north. (IAI)

Kfir C7 cockpit, the left side of the instrument panel (not seen in this picture) includes a digital display indicator. The navigation panel —which includes the INS— is located on the right console, besides the pilot's right leg. (IAI).

Meanwhile the Lavi program had attracted a lot of criticism based on the fact that the Kfir had been plagued with problems in its early days and had not entered service until a new generation of combat aircraft was already in existence. It was suggested that the Lavi fate would be the same. In 1987 101 Squadron was temporarily disbanded in anticipation of reactivating the 101 as an F-16C/D unit. This was the squadron first to dispense with the Kfir, being followed during the next years by the other squadrons, 109 converted to F-16C/Ds, 111 disbanded in 1994, 113 became an AH-64A Apache unit in 1990, 121 disbanded and 144 was reequipped with second-hand ex USAF F-16A/Bs in the summer of 1994. By mid-1995 the Kfir remained in service with five reserve squadrons, which in time of peace operate less aircraft with a core of professional air and ground crews. In case of exercises or war they are manned to their authorized strength, these are 104(R) and 132(R) at Nevatim, 143(R) and 251(R) at Ovda and 149(R) at Hatzor. These units will also be axed. By 1996 only three will remain and the Kfir will be phased out from Chel Ha'Avir service by the year 2000, or even earlier.

LIMITED EXPORT SUCCESS

Early in the Kfir program, IAI offered its fighter to potential foreign customers and almost a year after the first aircraft was handed over to the Chel Ha'Avir, IAI was negotiating the sale of the Kfir to six nations. Unfortunately for IAI, none of those deals materialized during the 1970s, mainly due to considerations that were irrelevant to the capabilities of the aircraft. They were mainly concerned with procurement cost and overall performance.

The Kfir was the cheapest of the Mach 2 Western origin fighters available during the mid-1970s, although not as cheap as the later Russian MiG-21 or Su-22 variants. As mentioned before, the Kfir C1 suffered from stability problems and poor maneuverability and Israeli pilots generally disliked the type, regarding it as an inferior fighter. IAI was quite sure it had a remedy for the Kfir problems in the shape of the canard and other aerodynamic refinements incorporated in the Kfir C2, while the problems of the C1 were kept secret at the time in order not to ruin the Kfir reputation worldwide.

Until the US Navy received its purpose built(?) F-16N and TF-16N for the Adversary role, the opportunity rose to loan a dozen Kfir C1s for VF-43 at Oceana, with the first reaching there in April 1985, two months later deliveries were completed. In Us service they were known as F-21A and dedicated to the adversary training role, simulating enemy threats as MiG-21s and MiG-23s. (IAI)

Compared with the F-15 or F-16, which were the Israeli pilots preference, the Kfir was indeed an outdated fighter, but when compared with fighters of its own generation, the Israeli fighter was in the top class. With a low price and high performance, Kfir was set to be an export success. In any event, it was not for three main reasons. The first was that generally it was offered to Third World nations that, although they desperately wanted to have a prestige Mach 2 fighter in their small air forces, were unable to absorb such an aircraft, operationally or economically. Thus, a number of negotiations such as those with Bolivia were fruitless from the outset. The second reason was that Kfir originated in Israel and was politically unacceptable to many nations. Such was the case in the Austrian contest for an economic interceptor, which technically and economically, was won by the Kfir during the mid-1970s, but not politically; the same happened with Taiwan. However, the third reason was the deadliest as far as Kfir exports were concerned, bad timing or simply bad luck.

This bad timing or luck stemmed from the fact that the Kfir is powered by an American powerplant and thus the U.S. Administration can veto any project sale. When the Kfir was revealed in 1975, Israel was in conflict with the Ford Administration over the pre-peace talks with Egypt and during the next year, when the C2 appeared, Jimmy Carter won the U.S. elections, marking the end of Kfir export sales in that decade. President's Carter philosophy was to restrict the export of sophisticated weapons to Third World nations. The Kfir, labelled in the brochures as a Mach 2+ multirole combat aircraft, was certainly viewed as such. Even a signed contract with Ecuador could not change the Carter Administration's mind and the foolishness of the policy was clearly demonstrated by the Ecuatorians, who simply ordered the Mirage F1 instead, although U.S. policy was changed shortly after the Reagan Administration took office in early 1980. Nevertheless, even during this period, U.S. considerations were not totally clear. In both Honduras and Mexico the Kfir was technically preferred but lost the competition to the F-5E/F which was backed by tempting American offers.

Finally, the Kfir obtained its first true export order when U.S. restrictions were eased and Ecuador ordered ten Kfir C2s, two TC2s, later supplemented by a third two-seater plus a simulator. These were assigned to Escuadrón de Combate 2113 "Leones," part of Grupo 211, Ala 21, at Base Aérea Taura, near Guayaquil. The first batch, delivered by sea, arrived in Ecuador on 25 March 1982. Assembled with Israeli technical support, the first Ecuadorean Kfir made its maiden flight at its new home on 19 April 1982 and the work up period for Escuadrón 2113 commenced. That year the Fuerza Aérea Ecuatoriana (FAE) launched the Taura Cup, an annual competition between its three supersonic combat squadrons (the other two, also based at Taura and forming part of Ala 21 are equipped with Mirage F1s and Jaguar Internationals). From 1985 until the end of the decade, the "Lions" won the cup every year. Ecuatorian Kfirs have been involved in combat operations at least twice against the Peruvians, as is related in the OTHER BATTLES Chapter. All surviving C2s were upgraded to C7 standards during the late eighties.

The Fuerza Aérea Colombiana (FAC) had long desired the Kfir. In 1981 it was reported that twelve Kfirs had been ordered to be delivered from March 1982, but that year the contract was canceled due to lack of cash. Nevertheless Colombian interest in the Israeli fighter resurfaced during 1986. In October 1987 the U.S. State Department withdrew its objections and by April 1988 the Colombian parliament approved the acquisition of 13 Kfirs. The aircraft comprised a dozen refurbished ex-Chel Ha'Avir C2s, which were brought to C7 standard in Colombia, and a single TC7, originally a TC2 but updated in Israel before delivery. Deliveries to Escuadrón 213 at BAM 5 "German Olano," Palanquero commenced in April 1989 and were completed by August of that year, with the TC7 delivered in early 1991. Also IAI provided support for upgrading the Dassault Mirage 5s in service with Escuadrón 212 at the same base, including avionics, canard winglets and air-to-air refuelling probes.

After the phase-out from Chel Ha'Avir service of the early Kfir C1s, remaining examples were stored at Hatzor air base, while IAI was looking for a potential customer. By 1984 a customer was

Two F-21As from VF-43 while on detachment to NAS Fallon, return home after a DACM hop. US Navy Kfir C1s were painted in two shades of grey. (US Navy)

VF-43 "Challengers" adversary fleet, from top to bottom, T-2C, TA-4J, A-4F and F-21A. In April 1988 VF-43 Kfirs were withdrawn from use and returned to Israel. (IAI)

Another Kfir C1s were leased to reequip a USMC adversary squadron, VMFT-401 "Snipers" at Yuma, Arizona, first deliveries were in July 1987, while retirement took place in September 1989, ending the American chapter of the Kfir. (IAI)

indeed found in the shape of the U.S. Navy that required a gap-filler solution for its adversary units until the first F-16Ns were delivered. IAI's proposition was tempting. The loan of Mach 2 fighters that could emulate the MiG-21 or MiG-23, with the payment restricted to maintenance only. The Israeli offer was accepted and an initial batch of a dozen C1s was ordered for delivery to VF-43 at Oceana NAS, Virginia in 1985. The Kfirs were refurbished by IAI before delivery while the U.S. Navy pilots converted to the type with the Chel Ha'Avir with 111 Squadron at Ovda in the Negev. The first three were delivered by sea in April 1985 with completion of deliveries two months later. That year negotiations commenced for the lease of another 13 C1s to equip the U.S.M.C.'s adversary unit VMFT-401 at Yuma, Arizona. An agreement was reached by 1986, the first eight were shipped to the U.S. during July 1987. In U.S. service the Kfir C1 received the designation F-21A, and a reputation as a rugged fighter. In spite of a very high annual utilization of some 300 flight hours per year, availability rate of the F-21As, maintained by IAS (a U.S. firm set up for the task by IAI), was impressive with a figure of near 95 per cent. Nevertheless, when the F-16N became available to the U.S. Navy, the F-21As of VF-43 were withdrawn from use in April 1988 and returned to Israel. During September 1989, the retirement of VMFT-401 F-21As, which were replaced by ex-USAF F-5E/Fs, ended the American chapter of the Kfir. All those returned to Israel were stored at the IAI works awaiting disposal.

Meanwhile IAI had developed the Kfir C10/TC10 advanced variant with latest generation avionics, although with the influx of new and later second hand F-16 Fighting Falcons, the Chel Ha'Avir was no longer interested in any upgrade on its Kfirs, which it had

USMC Kfirs carried several camouflage schemes, including the three illustrated in this picture, lizard three tone, lizard two tone, and air superiority grey. (IAI)

Kfir C2 patch.

The Technolog (ex Mirage IIIBJ 988) has been used during most of its life as a test platform, and served to clear the airframe/J79 engine mating, as well as to clear the canard foreplanes for operational use.

While in service with 101 Squadron Kfir C2 874 achieved the honor of obtaining the type's only air-to-air kill in IDFAF service, when on 27 June 1979 this machine shot down a Syrian MiG-21 with a RAFAEL Shafrir II AAM. This particular Kfir now rests at the IDFAF Museum at Hatzerim air base in company of other veterans. (author)

planned to phase out from service by the year 2000. The numbers stored is growing, and IAI salesmen have tried to seek clients around the world. In 1988 a contract was signed with the government of South Africa for the delivery of 45 Kfir airframes for conversion to Atlas Cheetah standards, including the replacement of the J79 for the Atar 9K-50 engine. For more details the reader can see the chapter about South African Mirages.

KFIR C7 SPECIFICATIONS

Wing: span 26 ft. 11.6 in.; canard foreplanes span 12 ft. 3 in.; wing area 374.60 sq. ft.; canard foreplane area 17.87 sq. ft.
Fuselage and tail: length 51 ft. 4.25 in. including probe; height 14 ft. 11.25 in.; wheel track 10 ft. 6 in.; wheel base 15 ft. 11.7 in.
Weights: empty about 16,100 lb.; normal take-off 22,961 lb.; maximum take-off 36,375 lb.
Fuel and load: internal fuel 5,670 lb.; external fuel up to 8,216 lb. in three 449, 343, 218, 159 or 132 U.S. gal. drop tanks; maximum ordnance 13,410 lb.
Powerplant: One IAI Bedek Division-built General Electric J79-J1E rated at 11,890 dry and 18,750 lb. with afterburning.
Speed: maximum level speed clean at 36,000 ft. more than 1,320 knots or at sea level 750 knots.
Range: ferry range 1,744 nm with one 343 U.S. gal. and two 449-U.S. gal. drop tanks; combat radius 419 nm on a hi-hi-hi interception mission with two Shafrir AAMs, one 218-U.S. gal. and two 343-U.S. gal. drop tanks, or 475 nm on a 1-hour CAP with two Shafrir AAMs, one 449-U.S. gal. and two 343-U.S. gal. drop tanks, or 640 nm on a hi-lo-hi attack mission with two 800-lb. and two 400-lb. bombs, two Shafrir AAMs and one 343-U.S. gal. and two 449 U.S. gal. drop tanks.
Performance: maximum rate of climb at sea level 45,920 ft. per minute; climb to 50,000 ft. in 5 minutes 15 seconds with full internal fuel and two Sahfrir AAMs; zoom climb ceiling 75,000 ft.; stabilized supersonic ceiling 58,000 ft.; take-off run 4,500 ft. at maximum take-off weight; landing run 4,200 ft. at 25,000 lb.; g limits +7.5.

KNOWN CHEL HA'AVIR MIRAGE/NESHER/KFIR SERIALS

A total of 76 Mirage IIIs, 72 IIICJs and four BJs were delivered between 1962 and 1986, while at least two of the single seaters were converted for reconnaissance with camera noses. These were operated by 101, 117 and 119 Sqns. The first to cease Mirage III operations was 119 Sqn. in 1971 followed by 101 Sqn. during 1975 and 117 during 1979. The remaining aircraft were operated by 190 Squadron since 1978 until phased out by late 1982. Most of the Mirages were reserialed more than once, as number 33 which later was 133, 733 and finally 833.
During the early period the Mirages were silver with a red and white stripped rudder (101 Sqn.), a red line along the fuselage (117 Sqn.) and a red lightning on the vertical fin. (119 Sqn.)

Known s/n:				
IIIBJ	286	87/287/787	88/988	289/789
IIICJ	02	03/103/703	06/406	107/507
	08	109/409	11/111	12
	14	15/915	118	19/719
		20/620/720	22/522	524725
	26/126	129/729	130/730	31
	732	33/133/733/833	34/534	741/941
	42/942	743	144	745
	147	148/748/948	649/749	150
	151/951	52/952	53/153/753	755
	56/756	158/458/758	159/259/459	60
	764	766	67	68/768
	70	171/771	775	76/176/776
	177/777	178/778	779	188/780
	81	82	83	84
	296			

A Total of 51 IAI Nesher S single seaters and ten Nesher T two seaters were built, known serial numbers are 501, 507, 510, 524, 525, 526, 527, 528, 531, 533, 534, 536, 537, 538, 539, 541, 542, 543, 549, 562, 563, 565, 566, 569, 569, 572, 572, 574, 582, 586, 598, 599, 701, 705, 707, 709, 710, 712 (converted as Kfir prototype), 714, 716 and 720.

212 Kfirs were produced by IAI, 40 C1s, 160 C2/C7s and 12 TC2/TC7s, some of the serial numbers used were taken from retired Neshers or Mirages: 703, 709, 710, 712, 714, 716, 724 to 728, 731 to 735, 739, 742, 747, 749, 750, 764, 785 to 787, 791 and 794 (C1s).

514, 517, 519, 523, 529, 534, 539, 555, 557, 705 to 708, 711, 718, 719, 730, 733, 736, 737, 738, 740, 743, 745, 748, 755, 759, 761, 779, 781, 804, 805, 808, 812, 813, 814, 821, 822, 824, 826, 827, 829, 830, 831, 833, 837, 838, 841, 843, 845, 846, 849, 851, 852, 853, 854, 855, 857, 858, 860, 862 to 868, 870, 871, 874, 877 to 880,, 882, to 885, 887, 888, 889. 891. 894 to 898, 903, 904, 910, 920, 928, 948, 961, 979, 981 and 987.(C2/C7s)

303, 304, 306, 308, 309, 310, 611 (TC2/TC7)

Ceremony at Yuma when the first F-21A was received for VMFT-401. (US Navy)

Ecuador received ten Kfir C2s and three TC2s, since then upgraded to C7/TC7 standards, equip Escuadrón de Combate 2113 "Leones", having entered in combat against the neighboring Peruvians at least twice. (Steinemann)

The other Latin-American Kfir user is Colombia which has acquired 13 Kfirs, twelve C7s and a single TC7, equipping Escuadrón 213, included in the deal were RAFAEL Phyton III AAM, as can be seen in this fine air-to-air study of this C7 carrying a Phyton III acquisition round. (Steinemann)

A highly modernized Kfir derivative was the proposed IAI Nammer, although this project did not see the light, the recent unveiled Cheetah C from South Africa bears an extraordinary resemblance, in fact South Africa acquired at least 45 Kfir airframes for conversion to Cheetah C and D standards. (IAI)

6

Mirages in the South Atlantic
Dagger Fighter Bombers

The first Daggers arrived in Argentina in December 1978 aboard the merchant vessel "Jaspers", being unloaded in the port of Buenos Aires. These jets were in fact, IAI Neshers constructed in Israel in 1974, and in accordance with the Dagger contract, 26 aircraft were supplied in a record time of six months, which also included a refurbishment, which tooK that name after entering in service. Additionally, in the Fuerza Aérea Argentina the aircraft were also known as Mirage M-5 Dagger. During 1980 the Dagger II contract was signed, which included 13 additional examples, plus spares, engines and a simulator. Mayor Sapolski was the first Argentine pilot to fly solo in a Nesher/Dagger, on 9 November 1978, while performing the conversion in Israel. These 39 machines were broken down as 35 Dagger A (Nesher A) single seaters, serialled C-401 to C-424; C-427 to C-437, and four Dagger B (Nesher T) two-seaters, C-425, 426, 438 and 439. The first Daggers concentrated at Mariano Moreno air base, until relocating to Area Material Río Cuarto (Air Depot) early in 1979. Meanwhile the Dagger's first public appearance took place at the Escuela de Aviación Militar (Military Air School) in January as Brigadier General Garaffigna took over the job of Chief of Air Staff.

The Escuadrón Aeromovil núm. 1 was activated at Rio Grande Naval Air Station in Tierra del Fuego, its C.O. was Major Carlos Napoleon "Napo" Martínez, a very experienced fighter pilot. This photograph, taken in October 1982, just four months after the war ended, at Rio Gallegos air base, where Grupo 6 maintained a detachment, shows Dagger C-414 inside a hardened aircraft shelter (HAS) in a state of readiness, loaded with four MK.82s and two 1.700-liter drop tanks. Posing besides the aircraft is Major Martínez.

On 8 August of the same year, the first Daggers were delivered to Tandil air base —this airfield which received the name of the city located nearby— is about 300 km south of the nation's capital, and also in the province of Buenos Aires. During the next days all available aircraft at Rio Cuarto were ferried to Tandil, and on 3 December, VI Brigada Aérea was commissioned, while the provisional Escuadrón Dagger became Grupo 6 de Caza, under the Air Defense Command (Comando Aéreo de Defensa). With the arrival of the second batch, more pilots were posted to the unit, which for operational purposes was divided into two squadrons.

After the Falklands War, and with the limitations of these aircraft in mind, as they had been designed for a completely different theatre of operations than that of the Falklands, the Fuerza Aérea developed an upgraded program which included a complete series of modifications, thus the surviving Dagger As received the avionics of the Kfir C7, including the navigation/attack computer, ranging radar, HAD and ECG. The retrofit was entirely made at the Area Material Rio Cuarto, with the support of several Israeli companies headed by IAI. Once upgraded the aircraft again changed their name, this time to Finger. In fact, initially the Nesher was a clear weather air superiority fighter, and after modification to Finger standards became an effective attack/interdictor, capable of operatlng In bad weather. Actually the Finger armament, for the air-to-air role, is based on the m DEAF cannon and the Sahfrir IV IR-homing AAMs, having seven hard points for external loads, against five in the Mirage, maximum warload for short range missions is established in 4.000 kg. Both jets, the Mirage and Dagger/Finger are powered by the same engine, the Atar 9C, the first built in France, the second in Israel. There are small differences between both engines, being only the engine controls, but are totally interchangeable.

The first Daggers were painted originally in a camouflage scheme similar to the Mirages. Later, like their brothers, they did lose the nationality markings and serials. Several experiments were made with new camouflage schemes, including light green and medium earth. But definitively during the late eighties they started to be painted in a medium grey sir superiority scheme. in mid-1995 Grupo 6 de Caza was formed by two Finger/Dagger squadrons, Escuadrón II and Escuadrón III, with some 22 surviving examples.

But now let's go back several years, to the eve of the Falklands war On 31 March 1982, Grupo 6 de Caza was under the command of Comodoro (Colonel) Thomas Rodriguez and the two flying squadrons in which the Grupo was divided were majors Juan

The amphibious assault ship HMS Fearless is near missed by a MK.83 bomb dropped by a Dagger at San Carlos Water on 23 May 1982. During the British amphibious landings Grupo 6 performed many strike missions in what was later called the Battle of San Carlos, which lasted from 21 to 27 May 1982. (MoD)

Grupo 6 pilots and ground technicians posing at Rio Grande in June 1982, note iced parking area; many missions were canceled owing to bad weather. (Grupo 6)

Sapolski and Carlos Napoleon "Napo" Martinez — both had made the Dagger conversion in Israel four years earlier.

The Argentine occupation of the Falklands Islands on 2 April 1982, came for the majority of Grupo 6 members as a big surprise; pleasant for some, worrying for the majority. First Lieutenant Cesar Roman, a Dagger pilot, explains his feelings then:""...As a flight commander I was given notice by my superiors of the Argentine landing close to Stanley a few hours before they took place, I was very nervous and tense by these events. As a member of a fighter outfit I had a great desire to participate in the hostilities, as that would be our baptism by fire, putting into good use all the daily training...."

On 2 April 1982, Grupo 6 had a strength of 37 aircraft (a single and a dual seat had been lost in separate accidents) of which about 60-70% were serviceable. Concerning the pilots, most of them did not have a lot of experience on that aircraft type due to the unit's recent activation, but there was a small cadre which were very experienced in fast jet operations (Skyhawk, Sabre and Mirage III). Summing-up, the group's readiness level was lower than desired in employing the Dagger as a weapons system, as one of its pilots, Jose Luis Gabari Zoco, said: "...I had just been posted to Grupo 6 de Caza, before, I was flying the A-4C at Mendoza. During early April I was instructed in the theoretical stage prior to flying the Dagger and I made my first flight on 20 April together with other newcomers; Captain Demierre and First Lieutenant Antonietti; up until June 1st, I flew as much as possible, flying replacement airplanes to San Julian and Rio Grande..."

For several of the Dagger missions, they were accompanied by the Learjets of Grupo 2 which with their complete set of navigation equipment acted as pathfinder until reaching near the western shores of West Falkland Island. (Grupo 6)

Also Cesar Roman had this to say: "....During most of April I got an intensive training at Comodoro Rivadavia in anti-shipping operations against target ships provided by our navy and one thing was clear, attacking a moving object as a vessel was a much more difficult target than I thought before..."

From the first day, Grupo 6 commenced an intensive training schedule in maritime strike and air combat. On 25 April, part of the unit's strength was deployed to southern Argentina airfields and thus two "mobile squadrons" were activated, number 1 based at Rio Grande naval air station in Tierra del Fuego with an initial strength of nine Daggers —also known as Escuadrón Dagger Rio Grande— commanded by Martinez, and number 2 based at the civilian airfield of San Julian in the province of Santa Cruz, Patagonia, with ten Daggers in strength, —also known as Escuadrón Dagger San Julian— commanded by Sapolski. Both squadrons were declared operational the same day.

In the air-to-air arena, the Dagger's main mission would be that of providing fighter cover to Argentine strike aircraft, as well as air superiority. Cesar Roman provides some details of the deployment: ""...The distance from Tandil to San Julian was covered in one hour and thirty minutes at a cruise speed of 500 knots

Escuadrón Aeromovil num. 2 was based at the civilian airfield of San Julian in the province of Santa Cruz, Patagonia, operating initially with ten Daggers and the same number of combat ready pilots which are seen posing in mid-May besides C-412, note the ship's silhouette besides the serial denoting a "share" on the 1 May attack against Royal Navy warships. (Grupo 6)

and at 35,000 ft. During the last week of April the tension among us reached a high level as the British Task Force continued due south and we were sure that the confrontation was unavoidable. In fact at 19:30 hrs on 29 April, the radar detected some "blips" far into the ocean approaching San Julian (probably caused by British countermeasures), the order was given for Daggers to re-deploy to Comodoro Rivadavia. The first aircraft were launched thirty minutes later and in ten minutes all were in the air —including an A-4C squadron—. The alarm proved to be false and by mid-morning on the next day we returned to San Julian..."

The Fuerza Aérea del Sur (Souther Air Command) second combat mission was the first for the Daggers when a section from the Rio Grande based squadron was launched at 07:45 hrs on 1 May for an air cover sortie over the Falklands, the configuration for this mission was two Shafrir II AAMs, 250 20mm HEI (High explosive incendiary) rounds and three 1.300-liter drop tanks. This section under the call-sign "Toro" was crewed by Captain Moreno as leader and Lt Volponi as wingmen. When at about 50 n.m. west of Port Stanley and at 22,000 ft the FAA (Fuerza Aérea Argentina) radar controller (a Westinghouse AN/TPS-43E portable set) based near Stanley advised about the presence of large numbers of Sea Harriers which just had raided the airfield. Two of the Sea Harriers went head-on for the Grupo 6 Daggers, they were LtCdr Kent and Lt Haigh from 801 Squadron, both sections crossed over, with the Daggers about 4.000 ft higher. They jettisoned the external tanks in order to have more agility in a dogfight, although neither the British, nor the Argentine pilots visually detected each other and the four fighters started to fly in circles trying to find out where the enemy was. The diminishing fuel level started to seriously worry the Grupo 6 pilots, which after some rapid mental calculation, understood that the better plan was to return home, so they turned west and pushed the throttles to full afterburner climbing as fast as the aircraft could go, reaching 37,000 ft, the best altitude for economical cruise. In spite of bad weather at Rio Grande, both aircraft, after an ILS approach, landed at this airfield without incident.

It is interesting to note, that in this (bloodless) air engagement of the war, both Argentine and British pilots, thought that each other had made use of air-to-air missiles, which was not the case, and this impression was caused by the drop tanks tumbling earthward trailing plumes of vaporized fuel.

At about 09:00 hrs two Daggers were launched from Río Grande with the call sign "Foco" crewed by Captain Mir Gonzalez and Lt Bernhardt, which were tasked to provide cover to several formations of Skyhawks from Grupos 4 and 5. They also engaged

This photo was taken in July 1980 at Tandil air base, home of Grupo 6 and shows from left to right, Lts. Bean, Bernhardt, Volponi and Aguirre Faget after their first solo flight in a Dagger, only Aguirre Faget would survive the war. (Grupo 6).

the Sea Harriers in a similar way as "Toro" section and again they had to break contact and return home, which was accomplished without incident but with very little fuel left in the tanks.

By noon, the FAS (Fuerza Aérea del Sur) headquarters issued a frag order for a further air cover sortie which was assigned to "Ciclon" section which found the skies over the Falklands empty of Sea Harriers.

During the afternoon's early hours, Brigadier General Crespo —FAS Commander— received information that several Royal Navy ships had been detected southeast of Port Stanley airfield, probably getting ready for a shore bombardment mission. Immediately he issued a fragmentary order to the air bases under his command to launch a new wave of strike and escort aircraft (the first one launched during the morning had not found any naval targets). At 15:30 and 15:40 hrs a pair of Dagger sections were launched from San Julian to provide cover for the fighter bombers with the call-signs "Dardo" and "Fierro" respectively, then followed "Torno" section composed of three Daggers armed with two 500-lb bombs and three 1.300-liter drop tanks each. Pilots of the latter were Captain Dimeglio as leader, Lt Aguirre as number two and 1st Lt Roman as number three. Five minutes later "Fortin" section was launched as escorts with Captain Donadille and Lt Senn as pilots. At 15:55 hrs "Rubio" section took off from Río Grande, comprising only one Dagger of two initially scheduled,

Three Grupo 6 Daggers on their dispersal at San Julian, note he yellow identification marks on the fins, wings and drop tanks. (Grupo 6)

its pilot 1st Lt Jose "Pepe" Ardiles did not abort the cover mission in spite of being justified to do so as his wingman developed a technical fault and had to abandon the take-off. It was a bad omen for the young Argentine pilot, cousin of a famous soccer star then playing in England. Captain Guillermo Posadas who was the "Escuadrón Aermovil No 1" maintenance officer at San Julian tells about his feelings during the afternoon's missions on May 1st "...Close to 15:45 nine Daggers were launched —which out of ten available at San Julian was a pretty good figure— seeing from the ground the departure of our friends in the fighter-bombers heading to the east was very moving, at the runway threshold with the Atar 9C turbojets at maximum power and engaging afterburners, the noise level was outstanding, the groundcrews waved to the pilots before take-off and these, like knights from a futuristic epic, with their sun visors down and oxygen masks fitted, returned the salute. After the noisy take-off, each of us who remained at San Julian began to wonder, would they come back? Would they drop their bombs and fire their guns? If they had to eject, would the ejection seat work? Truly, we were anguished and the feelings only dissipated when near the horizon line, over the sea we could discern the landing lights of the first Dagger approaching.

Lt. Aguirre Faget, deployed to San Julian, took part in the attack against three Royal Navy warships shelling Port Stanley airfield, HMS Glamorgan, Alacrity and Arrow, the Dagger pilots claimed they damaged two of them, although the only damage suffered was nine hits by 30mm shells on HMS Arrow. (Grupo 6).

For a short time a few Daggers operated from Comodoro Rivadavia airfield note the crowded parking area, with two Learjets, an IA-58 Pucará and a Bell 212 helicopter. (Grupo 6)

his Grupo 6 crew chief shows the damage caused to the intake cone of a)agger caused possibly by a rifle-caliber bullet or a shrapnel piece. (Grupo)

About one and a half hours later we received a message tation that our aircraft had engaged the enemy and that several 3ritish ships had been hit. We were very happy, mainly because ıll the Daggers were returning without damage. Sometime later, ıt dusk and after the last aircraft landed, we raced to meet the ›xcited pilots which provided us with first hand accounts of the ıttack, then we all went to the squadron's ready room for an ex-ensive post-mission briefing and to discover if there were any liscrepancies with the Daggers in order to get all ready for the ıext mission, the war had started in earnest..."

But let's return the main narrative to the solitary Dagger erialled C 433 flown by Jose Ardiles, which once over East alkland was vectored by the Argentine GCI to a Sea Harrier AP which approached from the northeast, seconds later he "tally-ıoed" the Fleet Air Arm pair (he was spotted visually at the same ıme by the British pilots) the Dagger was at 33,000 ft while the 3ea Harriers were at 20,000 and the three closed head-on. Ardiles started a shallow dive and launched a Shafrir missile against the 3ea Harrier flown by Lt Martin Hale from No 800 Squadron, ap-arently and in spite of having been fired out of its parameters, he missile locked into its would be victim, which turned away rying to evade it. The Shafrir pursued the Sea Harrier until 5,000 losing its lock-on and veering away, possibly by exhausting the ocket's propellant, although it passed uncomfortably close to the 3ea Harrier. Meanwhile Ardiles started a hard climbing turn with ull afterburner in order to continue with an altitude advantage ver the STOVL interceptors, then Flt Lt Bertie Penfold from No 00 Squadron in Sea Harrier XZ455 made good use of Ardiles actical disadvantage, with his Dagger silhouetted against the sky roviding a high IR signature. He locked an AIM-9L and fired, the 3idewinder ran the three miles separating both aircraft in a few econds and went up the Dagger's tailpipe. A bright fireball fol-

A cold morning dawns on Rio Grande Naval Air Station, with two Daggers being readied for a combat sortie over the Falklands. (Grupo 6).

lowed, almost disintegrating the Grupo 6 fighter, some pieces fell on Lively Island, the unfortunate pilot did not eject. The War's first air-to-air kill had taken place at approximately 16:40 hrs.

At Port Stanley, the GCI controller heard the anguished cry of Ardiles —who at the last moment saw the AIM-9L smoke trail and tried, to no avail, to avoid it in a hard break— disappearing from the radio frequency while his aircraft's blip vanished from the radar screen. It had been an almost suicidal mission and the price paid was heavy, an aircraft and a pilot. Squadron commander Martinez should had ordered Ardiles not to take-off when his wingman aborted.

Almost at the same time that C-433 exploded, "Torno" flight was approaching the Falklands, but now let's see how this mission developed as related by Cesar Roman:

Lt. Carlos Julio Castillo —with orange flight jacket— is effusively greeted by his squadron mates after a mission. Unfortunately he was shot down and killed aboard Dagger C-410 by Lieutenant Smith from 800 Squadron near Pebble Islands on 24 May 1982. (Grupo 6).

On 21 May a Dagger formation flown by Captain Mir Gonzalez, Lt. Bernhardt and Capt Robles attacked the HMS Ardent near San Carlos Water, it also was attacked by Grupo 5 A-4B Skyhawks and Navy A-4Qs, sinking later in the day. (E.Martinez).

"......About 100 miles west of the islands we descended to about 30-50 ft over the sea in order to avoid enemy radars and Sea Harrier CAPs, we were coasting their northern shores, then "Torno-1" (Dimeglio) saw an objective on the sea and we changed heading to attack it, when we were near, with some embarrassment we found it was a solitary rock which protruded from the ocean on the northern mouth of the Estrecho de San Carlos (Falklands Sound).

After arriving at the area where the enemy ships were supposed to be, we found nothing, then Dimeglio decided to continue to follow the coast's contour for several minutes. Near Stanley we could see artillery fire from shore positions against the sea. After changing the heading we saw three big gray warships in arrow formation engaging shore targets with their guns. The leader assigned a target for each one and we accelerated descending still more to avoid being detected, I did not fire the guns during the run for fear of being discovered. AA fire was very light and I think we caught them off guard. I dropped the two 500-lb parachute retarded bombs into "my" ship and after making some avoiding action, I climbed over Islad Soledad (East Falklands). On the way home, we were checking each other on the radio and were very happy to be all in one piece and alive.

When we were at about 15,000 ft I saw an aircraft at two o'clock in a diving turn. I alerted the leader, informing him that it did not pose a threat as it was getting away from us. After the uneventful landing at San Julian, I was surprised to find that the aircraft I detected diving was no other than Dimeglio, which following orders from the GCI went low level in order to avoid a Sea Harrier CAP which was closing dangerously. Due to the differences in fuel consumption between "Torno-1" and "2" and me, and as I did not hear the GCI alert call, I continued in my optimum cruise height, while they had to jettison the empty centerline tanks to create less drag. In fact during the landing circuit, I gave them priority. This was one of the missions in which the Sea Harrier followed us during our escape but we were supported by "Fortín" section which distracted their attention, as it seemed they were going after Aguirre..."

The Type 21 frigates HMS Alacrity and Arrow and the County Class destroyed HMS Glamorgan were caught by surprise as Roman said, although fortunately for them, the six bombs fell into the water, two or three being near misses causing some structural damage to Glamorgan and Alacrity, while Arrow received eleven hits from 30mm guns on her superstructure near the funnel.

During the "Torno" section escape, they were pursued for about 80 miles by a Sea Harrier section. As the AAM-armed "Fortin" section was nearby, it was directed to intercept the Royal Navy fighters. Low on fuel and with "bogeys" approaching in a stern conversion, the Sea Harriers split-essed and abandoned the chase.

These were the last sorties flown by Grupo 6 on May 1st and the events showed the Dagger pilots their insufficient fuel capacity to establish a CAP over East Falkland island for more than three/five minutes, and that without any margin for mistakes in computing the best approach and egress profiles. Another element was the differences among the Shafrir II and the "Lima" and although the former was not a bad missile at all, the AIM-9L was much more advanced, but most important of all was the big difference in air combat training between the Sea Harrier and the Grupo 6 pilots. The former were then, and still are, among the best in the realms of modern air combat, and were not lured away to engage the Daggers at high altitudes, where the Sea Harrier advantages somewhat diminished.

A Dagger and its pilot had been lost in a courageous but irreversible act, engaging in combat without the support of a wingman, something vital in the air-to-air arena. From then on the Daggers were tasked almost exclusively with strike missions, mainly against British ships.

On 4 May 1982, two Dagger sections from Escuadrón Aeromóvil No 1, were launched at 10:20 and 15:00 hours respectively from Rio Grande to provide escort for two Grupo 1 de Transporte KC-130H Hercules tankers which refueled the pair of

The Westinghouse AN/TPS-43 mobile radar installed by the Argentine Air Force near Port Stanley provided valuable information to Dagger and Mirage formations over the islands. (author collection).

Pilots and ground crew assists to a catholic mass inside a shelter at Rio Grande. (Grupo 6).

Aviación Naval Super Etendards on their attack to HMS Sheffield. The used the call signs "Tango" (Robles/Cimatti) and "Mate" (Dimeglio/Mir Gonzalez), the mission developed uneventfully.

During the following days weather conditions were very bad and no missions were flown but on 9 May the skies cleared somewhat. Accordingly two fragmentary orders were issued to San Julian and Rio Grande respectively, these indicated "possible" naval targets near Port Stanley. The first to launch was "Puma" flight formed by four Daggers which did it at 13:46 hrs followed at 15:09 hrs by "Jaguar" flight with the same number of aircraft. Unfortunately for the Argentine pilots, a few minutes before arriving to Jason Island (NNW of West Falkland) they had to turn back due to very low cloud cover with gusting winds and rain/snow showers.

Three days later, on 12 May, a combined attack was planned with two flights of A-4Bs from Grupo 5 de Caza and another two of Daggers to attack Royal Navy warships which were engaged

Artist impression of the 1 May 1982 attack against three RN frigates by Torno flight —Capt. Dimeglio, 1st Lt. Roman and Lt. Aguirre—. The true fact is that only HMS Arrow was slightly damaged by 30mm gunfire. (E.Martinez).

This painting shows the rescue operation performed on 29 May 1982 by a DHC-6 Twin Otter from Grupo 9 at Calderon airstrip on Pebble Island, returning to the mainland were two Dagger pilots which had been shot down the previous days, Major Puga and Captain Díaz, the later wounded, as well as the remains of Lt. Volponi. (E.Martinez)

in shore bombardment duties, but due to technical and communications problems, it was impossible for the Grupo 6 aircraft to take part in the mission, with the Skyhawks going alone and incidentally suffered heavy losses.

The days passed with some alerts and a few unproductive local air defence scrambles, mainly caused by radar contacts coming from inside the Chilean territory in Tierra del Fuego —at that time the FAA staff feared a Chilean preemptive raid in order to solve in their favor the Beagle Channel dispute— other evidence showed that a pair of RAF ELINT Camberras were secretly deployed to Punta Arenas.

The evening of May 20 was very tense for the Grupo 6 personnel stationed at San Julian and Rio Grande, all kinds of rumors were afoot, about possible British landings on the islands, it was even feared that a possible SAS/SBS commando raid to Rio Grande to destroy the Argentine Navy Super Etendards might take place. Most of the pilots had a hard time trying to sleep that night, the eve of dramatic events for all them.

21 May dawned cold and humid but with clear skies, on the Falklands the weather was similar, although a front was approaching.

At Comodoro Rivadavia air base, Headquarters of the FAS, Brigadier General Ernesto Horacio Crespo, its C.O. was nervous and impatient, he was up since 05:00 hrs in the early morning and had not abandoned since then the war room, he was trying to receive reliable information from the islands as the last of his thoughts was to send his "boys" against phantom targets. Then at about 08:30 hrs sketchy information arrived about supposed British landings at the north of the Falklands Sound but without any confirmation. Two hours later, definitive confirmation was received that it was not a feint, the British were establishing a beachhead at San Carlos, but some time before, he had taken the decision to send some of his aircraft in armed reconnaissance missions to find and attack the British ships. "Ñandu" flight was launched at 09:44 hrs from Rio Grande, its pilots were Captain

This picture of a C-415 taken after the war shows the aircraft loaded with Mk.82 bombs, 500-liter supersonic tanks and inert Shafrir AAMs. (Grupo 6)

Rodhe, Lt Bean and Captain Janett. "Perro" flight departed one minute later and was formed by Captain Moreno, Lt Volponi and Major Martinez. Each Dagger was armed with a single Mk.l7 1.00-lb bomb and two 1.700-liter drop tanks. Captain Diaz, Lt Aguirre and Captain Dellepiane were launched at 09:53 hrs from San Julian with the call-sign "Zorro", while Captain Dimeglio and Lt Castillo were "Leon". At 10:25 hrs the first Daggers arrived to the San Carlos area, "Ñandu" flight attacked the frigates Broadsword and Argonaut, only the former suffered damage in the form of several hits by 30mm gunfire from Bean and Janett, however a Sea Wolf missile launched from the Type 22 frigate, bagged Bean's Dagger, he was killed as C-428 disintegrated just above the water. "Perro" flight was the next in the attack and went for the County Class destroyer HMS Antrim, which received a direct hit; the Mk.17 bomb went through several compartments coming to rest without exploding in the Sea Slug missiles storage room; also the unfortunate warship received several 30mm rounds. Due to this damage she was temporarily out of action, seeking refuge in San Carlos Water to make emergency repairs and deactivate the bomb. Meanwhile the three Daggers escaped unscathed from the inferno of AA and missile fire, overflying the sound on a southerly course. During this phase, one of the Daggers was so lucky that a Sidewinder launched at extreme range by Lt Martin Hale

The original insignia of Grupo 6 de Caza. (Padín)

from No 800 Squadron did not hit the Grupo 6 fighter as the missile exhausted its rocket propellant and fell off, Hale and his element leader tried to get into range but without drop tanks, the Daggers were

going extremely fast and the Sea Harriers had to abandon the pursuit due to fuel shortage.

At 10:51 "Zorro" flight made its appearance in the battle area, attacking the frigate Brilliant and although an electrical failure prevented the bombs being dropped, the Grupo 6 pilots emptied the 30mm guns magazines into the Type 22 frigate, which received at least twenty impacts. The operations room and the wardroom were affected, causing several wounded among the crew. Two minutes later "Leon" section attacked two ships with bombs and cannon without causing any damage.

The third wave of Argentine aircraft (the second was exclusively composed by A-4B/C Skyhawks) began to take-off during the afternoon first hours, thus "Cueca" flight was launched from Rio Grande at 15:55 hrs being formed by Captain Mir Gonzalez, Lt Bernhardt, Captain Robles and 1st Lt Luna. Five minutes later "Laucha" flight was launched from San Julian with 1st Lt Roman, Major Puga and 1st Lt Callejo as pilots, and a few minutes later it

Patriotic image of an upgraded Dagger in landing configuration and framed against the Argentine flag. (Grupo 6)

After the war, surviving Daggers were upgraded to Finger standards, with a new nav/attack system based upon that of the Kfir C7, including its nose cone for the Elta ranging radar. (Padin).

The first Daggers (ex-IAI Neshers) lined-up at Ben Gurion International Airport, prior to delivery to the Argentine Air Force. (IAI).

was the turn of "Raton" flight also from the same airfield with Captain Donadille, Major Piuma and 1st Lt Senn.

The four Daggers from "Cueca" flight now flying at very low level, got "feet dry" on the islands heading east through a valley formed by Mt Marla and Mt Robinson of the Morandy chain. The weather conditions forced them to fly lower than the hills with some of them covered by a deep cloud blanket, they flew in trail formation —the only possible way in these conditions. The leader discerned between the mountains located to both sides, something resembling a hole between the cloud cover and the valley; through it went the four Daggers. They were roughly north of Chartres Settlement when the Argentines were detected by radar while a Sea Harrier CAP from No 800 Squadron lost no time going after them. Robles and Luna (Nos 3 and 4 respectively) were visually detected by LtCdr Frederiksen aboard Sea Harrier XA455, the Royal Navy pilots firewalled the throttles and approached to launching distance of the AIM-9L. With the growl tone growing on his headset, Frederiksen pressed the button, the missile on the left pylon went for "Cueca-4", just before impact, the Grupo 6 pilot saw through the corner of his eye the Sea Harrier flown by Sub Lt George (Frederiksen wingman) with urgency, he tried to alert his comrades...the events that followed are narrated first hand by Hector Luna:

".....A few seconds after seeing the Sea Harrier and when I was trying desperately to alert the other members of the flight, I saw a fireball through the cockpit's rear view mirrors, immediately I felt the missile hit on my aircraft and it went out of control with the stick useless in my right hand. I tried to climb but to no avail, then the aircraft rolled inverted. I thought I was a dead man. Releasing the stick, I frantically searched for the upper face blind ejection handle. I believe that when I reached it, the aircraft was upright as the ejection was normal. I felt the explosion, a great noise, then a jolt and the chute opened almost at the same time I hit the ground very hard, breaking my left shoulder while getting a knee dislocated..."

Luna's Dagger, serialled C-409 had crashed —uncomfortably close to him— about 20 km west of Port Howard. Next day the wounded Grupo 6 pilot was found by a group of kelpers, which helped him contact by radio with the Argentine HQ at Port Stanley, being recovered by a Grupo 7 Bell 212 helicopter and flown to Goose Gree and after to Port Stanely. On May 25 he was evacuated to Comodoro Rivadavia in an C-130 Hercules.

Robles was unaware of what had happened, as well as that Frederiksen fired about 200 30mm rounds at him without hitting! Also Mir Gonzalez and Bernhardt were unaware that George tried to bag them, albeit unsuccessfully with 30mm gunfire. All three were too worried trying to avoid crashing while flying through the narrow valley. Once they found the exit, Robles broke radio silence to say that Luna had disappeared from behind him. They thought he had crashed against a mountain during the last stage of the dangerous low level navigation. The diminished "Cueca" flight crossed the eastern shore of West Falkland and were over

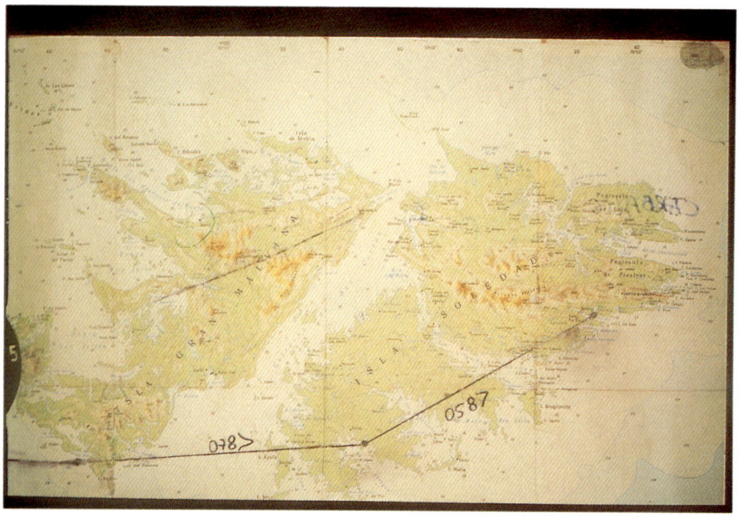

Map of the Falklands showing the route planned for some of the strikes planned for 8 June 1982. (Padin).

The insignia of Escuadrón Aeromovil núm. 1. (Padin)

the waters of the Sound, at about halfway they saw the frigate HMS Ardent at Gantham Sound, south of San Carlos. They made a little heading correction to attack her across the stern. The Daggers received light AA fire, when within effective range, they opened up with their guns and dropped the Mk.17s. Mir Gonzalez bomb fell about 12 meters short, making a big column of water, somersaulting and imbedding without exploding in the stern. Berhardt's also struck the stern, but this one exploded, destroying the Lynx helicopter, demolished the hangar and the Sea Cat launcher and caused several dead and wounded among the crew. Robles' bomb fell in the water without doing any harm, when overflying the frigate. This pilot saw pieces blown up into the air in a odd manner, he glanced at the clock in the instrument panel, it was 14:47 hrs.

The three members of "Cueca" flight egressed through the Falklands Sound on a southerly heading, and after almost one hour arrived back to Rio Grande without further trouble. HMS Ardent was badly wounded, although she still had some fighting power left, setting course to San Carlos Water in order to get the mutual support from other ships and to affect emergency repairs. Out of the six Daggers which took off from San Julian, the first to reach was "Laucha" flight, its attack is described here by Cesar Roman: "....The last phase of the flight was made bordering the northern coast of Gran Malvina Island (West Falkland). Once Rosalie Mountain was passed we found the Sound's waters and after flying very low, we found that we were higher than expected as the eastern shore is high and abrupt, we pushed the stick forward to "scrape" the waves and thus avoid the ship's radars. I saw a warship near San Carlos Water, some flames and a long column of smoke poured from it, the result of an attack by another Dagger flight. With number three man (Major Puga) I attacked a frigate which with her AA armament raised a wall of water, in fact I went through one of these geysers, which considerably reduced the visibility and the aiming. Together with number three I fired the cannon and dropped the bombs, then we egressed at full bore with an easterly heading, while number two attacked another warship which was located to the north of us. An anecdote of this flight was that after the attack, we had to change course just one minute after dropping the bombs in order to follow an escape corridor prepared beforehand during the mission's navigation planning. As I overflew the frigate, I triggered the chronometer (I had not in sight number two) and said over the

The insignia of Escuadrón Aeromovil núm. 2. (Padin)

One of the streets at Tandil air base was named in the honor of Captain Jose Luis Ardiles, a Grupo 6 pilot who was shot down and killed on 1 May 1982 by a Sea Harrier FRS.1 of No 800 Squadron flown by Lt. Martin Hale. (Padin)

After the war, surviving Daggers were upgraded to Finger standards, with a new nav/attack system based upon that of the Kfir C7, including its nose cone for the Elta ranging radar. (Padin).

radio: 40 seconds to go", so he could know the time left before changing heading. During the attack run Major Puga noted that something hit his aircraft, on the return leg while we were climbing to cruise height, I closed-in to observe his Dagger from several angles, but did not find nothing abnormal, after we landed a hit was observed on the front armored windscreen at about the same level as the pilot's helmet, but fortunately the bullet or shrapnel piece only dented the glass without piercing it. After the proper happiness to be back at home in one piece, we found that "Raton" flight return time was overdue, next day we learned that its three members had been downed by Sea Harriers, but all survived..."

Roman and Puga had attacked the frigate HMS Brilliant but their bombs missed, although she collected several 30mm rounds causing slight damage and causalities. Meanwhile Callejo attacked a vessel described as a frigate but his bombs exploded in the water. The three members of "Laucha" flight escaped pursued by a curtain of tracer fire and even some missiles.

One minute later followed "Raton" flight and its misfortunes are described with all detail elsewhere in this chapter, suffice to say that the flight was visually detected by LtCdr Ward (C.O. of No 801 Squadron) in Sea Harrier ZA195 and after a crossover and some maneuvering, Piuma in Dagger C-404 was downed by Thomas, then he went after Donadille in C.403 bagging him with his second and last Sidewinder. After a short pursuit Senn was shot out of the sky by an AIM-9L fired by Ward: "...The Dagger changed to a westerly heading, I was at 300 ft over the terrain, dived after him, got a lock-on and fired an AIM-9L. It could not fail! Seconds later it hit the target and, as if in slow motion, I watched as the Dagger disintegrated, hitting the ground, then the right engine intake and the wing's leading edge from the same side cartwheeled among smoke, flames and debris. I did not observe ejection, but later learned that he had done so and survived..." The three fortunate pilots from "Raton" flight were rescued by their own forces and carried to the relative security of Port Howard. With this engagement came to an end the missions of Grupo 6 de Caza during the dramatic journey of May 21. The Fuerza Aérea Argentina had launched 54 sorties from mainland bases (of 63 initially planned) losing seven aircraft, an alarming 13%. But summing up Grupo 6 activities, its losses were very heavy, 23 escort and attack sorties were accomplished and five Daggers did not return, which puts the sortie versus loss rate on a staggering 23,5%. The luck of the five pilots was not known then, but during the night, news reached Comodoro Rivadavia that four of them had been recovered.

Some factors had contributed to these high losses, first the lower experience level in the Dagger (compared to the A-4B and C), also the lack of a sound tactical background in anti-shipping operations and mutual support when bounced, but perhaps the worst mistake was that they went for the warships instead of the transports. In the positive side —although not much was known then— they had caused heavy damage to the frigate Ardent,

One of the original four Dagger B two seaters, seen while undergoing planned maintenance at one of Tandil air base hangars. (Padin)

moderate to the destroyer Antrim and light to other vessel. Some measures were studied during that night, specially the flight profiles over the islands in an effort to avoid the Sea Harrier CAPs.

On May 22 there were no Dagger sorties due to bad weather conditions. Next day it improved a bit and at 08:43 two flights were launched from Rio Grande which had to turn back due to very low clouds over the islands. During the afternoon, at about 15:00 hrs three sections of Daggers were over the Falklands: "Puñal —Major Martinez and Lt Volponi—, "Daga" —Captain Rodhe and 1st Lt Ratti— and "Coral" —Captain Dimeglio, Lt Aguirre and 1st Lt Roman— the first was from Rio Grande while the other two had been launched from San Julian. Alerted by the FAA radar controller that there were Sea Harrier CAPs nearby, "Puñal" dropped the bombs and turned home at low level; both Daggers were detected north of Pebble Island by LtCdr Auld C.O. of No 800 Squadron and his wingman, Lt Hale. Martinez was going too fast to be caught, but Volponi in C-437 had delayed in pushing the throttle to the firewall and was some ways behind. Hale in ZA194 was about two miles behind Volponi and when the distance had been reduced to half a mile, he launched a Sidewinder. The unfortunate Grupo 6 pilot never tried any evasive maneuver, the missile flew up the Dagger's tailpipe exploding in a brilliant flash, the doomed fighter bomber commenced to disintegrate cartwheeling into the ground west of Elephant Bay on Pebble Island, Volponi was killed.

On May 24 the FAS planned a total of 12 fragmentary orders for attacking British shipping in San Carlos Water. At 11:02 hrs the first Dagger flight arrived to the area, it had the call-sign "Azul" and was from Escuadrón Aeromovil Num. 1. Each Dagger carried two 1.700-liter drop tanks and one Mk.17 on the centerline station, number three of the formation was Captain Carlos A. Maffeis, here are his words of the attack:

"..."Azul" flight of which I formed part, was integrated by four Daggers having been briefed to attack naval targets within San Carlos Water. As I remember we were the first to arrive and in some way we "surprised" the British, perhaps this is why the four of us exited the area in one piece although not before getting hit by small arms fire. On that day I was flying aircraft C-431 and my place in the formation was as section leader (No 3), my wingman was Captain Robles and the flight leader and his wingman were

A Dagger seen on its landing run and with its cruciform braking parachute deployed. 35 Dagger As were delivered to Grupo 6 in two batches. (P.Steinemann)

Captain Mir Gonzalez and Lt Bernhardt respectively, this latter pilot was shot down some days later on this same place. As usual we ingressed at very low level and after we crossed the Bahía Ruiz Puente (Grantham Sound) and the nearby hills on a south to north radial, we entered San Carlos. Contrary to my three buddies I went over its eastern shore where there was a major accumulation of ships. On the west side there was a big transport, my decision to attack smaller vessels was based in that four minutes before, the generator of C-431 had failed and with this breakdown the bombs could not be dropped. It was a limitation of the aircraft's electrical circuit and I only had the possibility of firing the two 30mm guns. I pulled the trigger three times, the first impacts were short and raised big plumes of water, the next perhaps impacted on the target.

All happened very fast and as in other opportunities, it remains firmly recorded in my mind and the gun camera films, when these were developed, besides the technical details (350-degree attack heading), six ships could be seen, or about fifty percent of all British vessels inside San Carlos Water..."

"Azul" flight caused only light damage to the amphibious ships, nevertheless the four Daggers were profusely photographed by British photographers. Fifty minutes later "Plata" flight coming from Rio Grande, arrived in San Carlos, this time armed with two 500-

Map of the operations area showing the route of some of the missions flown by Escuadrón Aeromovil núm. 2 from their base of Rio Grande. (Padin)

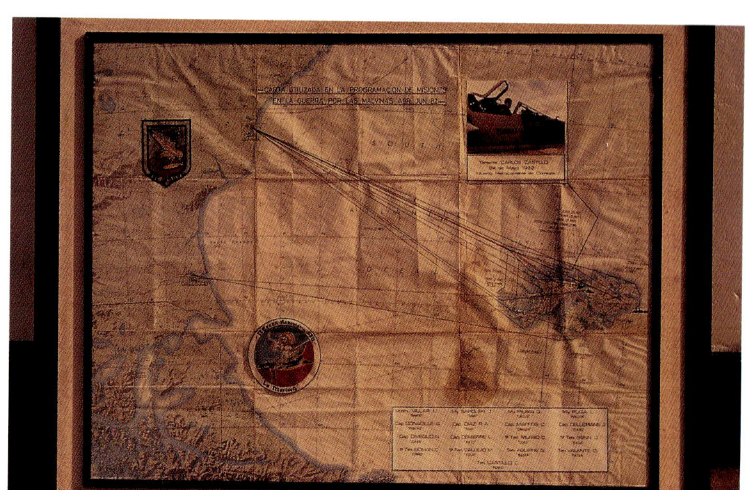

Map of the operations area showing the route of some of the missions flown by Escuadrón Aeromovil núm. 1 from San Julian, and Rio Gallegos. (Padin).

Two of the Daggers upgraded to Finger standards. Although it was initially planned to incorporate air-to-air refueling capability, this was later discarded due to financial constraints. (P.Steinemann)

lb parachute retarded bombs each; pilots were Captain Dellepiane, 1st Lt Musso and Lt Callejo. Once bombs were dropped over the targets, they egressed on a northerly heading, at ten miles from Pebble Island they found two warships at considerable distance, one fired to SAMs, which by their size thought to be Sea Darts. Fortunately for the Grupo 6 pilots, evasive maneuvering made them avoid the "white telephone poles" as they were nicknamed by the Argentines.

Some time later, a flight from San Julian, "Oro" with Major Puga, Captain Diaz and Lt Castillo was heading towards San Carlos following the northern coast of West Falkland to hide from radar. Before passing Pebble Island, the three Daggers were detected by the frigate HMS Broadsword which was acting as radar picket and advanced air defence ship (together with the Type 42 destroyer HMS Coventry). Quickly the ship's controller vectored a Sea Harrier CAP from No 800 Squadron flown by LtCdr Auld in XZ457 and Lt Smith in ZA193. The naval interceptors climbed to have a better view and to dive after them gaining speed in the process and negating the Dagger's speed advantage; the Argentine fighter bombers tried to escape jettisoning the external stores but the bounce was very well planned, Auld launched his two AIM-9Ls one after the other. The first hit Puga's Dagger C-419 which exploded and went into the water, the pilot was able to eject during the last seconds:

"...We were flying very low over the waves, we saw the Sea Harriers at the last moment and tried to evade them, but to no avail, then I saw Castillo's aircraft blowing-up and at the same time felt a missile hitting my aircraft. I tried to gain some height pulling back the stick, but I noted that the airplane started to roll, flames were pouring from the cockpit backwards, then a wing folded. I ejected with a 90-degree bank angle parallel to the water surface. I never lost consciousness and remember perfectly the jerk as the parachute opened, the first surprise was to feel alive, thanking god for it, the parachute canopy was not completely deployed when I hit the water...""

Puga's ordeal was not over because although he was not injured, he spent several hours in the cold water struggling to keep his head dry. He swam to Pebble Island and spent the night in a rocky beach. Next day he was found by an Argentine patrol led by a Marine Lieutenant. On May 29 he, together with Raul Diaz, several other personal and Lt Volponi's remains were picked-up in a daring rescue operation led by Captain Uriona of Grupo 9 in a DHC-6 Twin Otter. Puga's wristwatch stopped at 11:08 hrs, time of his ejection. To this day he keeps it as a reminder of these dramatic hours.

Auld's second Sidewinder got Captain Diaz in C-430, he ejected and the Dagger fell in the water to the north of Pebble Island, quite close to where Puga's aircraft fell. Smith fired an AIM-9L against Castillo in C-410 in a text-book lock-on, hitting the Dagger, which burst into flames and fell to the ground on the western side of Elephant Bay, the unfortunate Argentine pilot was killed. Both Sea Harriers pursued for a few seconds a fourth Dagger which they thought formed part of the same flight, but in fact it was number three of "Plata" flight which made good his escape.

Again Grupo 6 de Caza had suffered grievous losses caused by a deadly duo —Sea Harrier/AIM-9L—, with "Oro" flight completely wiped-out, all for scarce positive results. Certainly this loss rate could not be supported for much time, as the risk was that the Grupo ran out of aircraft. Six out of eight pilots downed by Sea Harriers during the Battle of San Carlos (21-27 May) had been saved thanks to the Martin Baker JM-6 zero-zero ejection seat.

This Dagger, photographed in April 1990, shows the markings of three British warships attacked and damaged. Argentine pilots made courageous, sometimes near suicidal, attacks against the British Task Force, having to pass through the Sea Harrier CAPs and the gauntlet of flak and SAMs. (Padin)

A Grupo 6 pilot, looking at his Argentine AF issue wristwatch. On the right sleeve of his flight jacket he carries the old Grupo 6 insignia and the Falklands campaign insignia, besides on the chest is the new Grupo 6 insignia; his neck scarf carries the colors of the unit, blue and dark red. (Padin)

Fitted with two 1.300-liter drop tanks, this Dagger is seen before departing for a training hop. (Padin)

On the following days, partly by bad weather and by a complete reorientation of the strategy to follow by Grupo 6, the Daggers made very few sorties as on May 25 in which two sections were launched from Rio Grande at 11:00 and 11:02 hrs respectively ("Ñandu" and "Fuego" on an offensive reconnaissance mission over the Falklands Sound southern side without finding nothing. Next day, a Dagger flight, "Llama", dive bombed through the cloud cover British troop concentrations near Port San Carlos, no AA fire was spotted by the pilots, neither the results achieved by the parachute retarded bombs. On May 27 the beachhead was definitively consolidated while British troops commenced to advance to other points. This period of six days was called later the Battle of San Carlos, being considered a victory for the forces of the United Kingdom in spite of the courageous, but in some way inefficacious efforts of the Argentine combat aircraft. Grupo 6 contributed with a total of 47 escort/cover and attack sorties accomplished (21 from San Julian and 26 from Rio Grande) losing nine aircraft and three pilots, getting in exchange what could be called as meager results, some Daggers had been hit by small arms fire or shrapnel, and although returning home, were temporarily out of service to affect repairs. On May 28, in order to support the Argentine garrison at Goose Green and Darwin which was hard-pressed by advancing British Paras, the FAS planned ten fragmentary orders, among them a Dagger flight. The mission was to attack advancing UK troops as well as the ships engaged in shore bombardment. Unfortunately for the Argentine ground forces, the weather conditions were not favorable for conducting air operations, the islands were affected by a low pressure area producing rain and snow precipitations together with very low visibility. The four Daggers could not employ their weapons, being forced to return home.

Left side view of another pilot showing the Mirage patch. (Padin)

Next day the weather improved a bit and the FAS ordered that several flights be launched. The first to arrive to the operations area at 12:25 hrs was "Limon" section with Captain Mir Gonzalez and Lt Bernhardt from Rio Grande, both were tasked with offensive reconnaissance and loaded with three 500-lb parachute retarded bombs. The pair arrived to San Carlos water and did not find any shipping there or in the Sound, so the leader thought that it was better to save the bombs for another mission and return home, but then Mir Gonzalez discerned to the north at extreme distance the silhouette of a large ship. Seconds later Juan Doming Bernhardt in C-436 who was some distance behind the leader was shot out of the sky by a Rapier missile, the Argentine pilot and his Dagger plunged into San Carlos Water.

Almost two hours later, "Puma" flight formed by two Daggers (a third had aborted) approached to the Falklands Sound by the north

of Pebble Island, then to the south they found a Sea Harrier CAP, believing that they were going to be intercepted, they jettisoned the stores and returned to San Julian. In fact it was an 800 Squadron section which was attacking the airfield of Pebble Island and their pilots had not seen the Daggers.

Bernhardt was the fifth and last pilot lost by Grupo 6 de Caza during the conflict, while C-436 was the last Dagger lost. All these losses in about nine days of effective fighting were a dramatic event which affected the unit's personnel, especially taking into account the overall results obtained.

Then followed several days in which the weather conditions hindered the normal mission tasking. On June 4 the situation of the Argentine garrison in the Falklands was beginning to get complicated, thus the FAS decided to launch several close support sorties, but the weather with low clouds (with a minimum ceiling

Starting late in 1989, all Daggers and Mirages commenced to receive the fashionable gray air superiority finish, accentuating more the Dagger's speedy lines. (Padin)

When the initial Mirage III interceptors were acquired, in the original lot were included two Mirage IIIDA two-seat conversion trainers, coded I-001 and I-002 (I stands for interceptor), 001 was w/o in March 1979 and 002 was seen in October 1982 at the former base of today's disbanded Grupo 8 de Caza, Doctor Mariano Moreno in the province of Buenos Aires. (Romero)

of 150-ft and a maximum of 500-ft) and rain squalls, made impossible the low level attacks, in turn it was decided to make medium height level or dive bombing attacks.

A sortie was planned for Escuadrón Aeromovil Num.2, the pilots assigned were Comodoro (Colonel) Villar, Captain Demierre, 1st Lt Musso and 1st Lt Roman. Each Dagger carried two 500-lb parachute retarded bombs. At 16:30 hrs the four aircraft were over the target area which was near Mt Kent, then dropped their bombs in 60-degree dives among the clouds, although results were not observed, all the Daggers returned home.

At 14:46 hrs next day "Fierro" section arrived over the island armed only with the 250 rounds of HEI of their 30mm DEAF cannon per each aircraft, making a reconnaissance run about the full length of the Falklands Sound without observing any British ship and returning home unscathed. At 15:30 hrs the three-Dagger "Nene" flight was over West Falkland tasked to make an offensive reconnaissance of King George Bay starting from the south inlet by inlet, creek by creek of its abrupt coast looking for a Type 42 destroyer presumably on radar picket duties, they could not locate their objective which apparently was to the south of West Falkland.

For a change in the cloudy or misty weather of past days June 8th dawned with crystal clear skies dotted with some clouds. In the FAS war room at Comodoro Rivadavia, messages started to pour since early in the morning from the islands, as a new beachhead was being established by the British at Port Pleasant about 27 km southwest of Port Stanley. Two amphibious ships were within the bay unloading supplies and equipment without any hurry, as in peacetime, and apparently without any defensive measures against possible air attacks, but Brigadier General Crespo was perfectly aware about the carnage the Sea Harriers could inflict upon his fighter bombers if they were not temporarily kept away from the Port Pleasant area. The fragmentary order was relatively simple and initially included Grupo 5, Grupo 6 and the Mirage III from Grupo 8 which should perform decoy duties simulating fighter-bomber flight profiles heading into San Carlos, also a Dagger section was tasked with this role.

The Grupo 6 attack mission was formed by six Daggers divided in two flights: "Perro" with Captain Rodhe, 1st Lt Gabari and 1st Lt Ratti; "Gato" with Captain Cimatti, Major Martinez and 1st Lt Antonietti. Both flights were launched at about 13:00 hrs from Rio Grande, each aircraft carried two 500-lb Mk.82s and full ammunition for the 30mm guns and two 1.700-liter drop tanks.

The first lot included ten Mirage IIIEA interceptors, serialed I-003 to I-012 equipped Grupo 8, formerly I Escuadrón de Caza Interceptora. I-008 was photographed while undergoing maintenance at Dr Mariano Moreno in October 1982. (Romero).

ust after take-off Cimatti collided with a bird causing some cracking on the windshield and thus having to abort the mission, the ive fighters rendezvoused with a Learjet coned TC-23 from Grupo Aerofotografico (Photographic Group), which acted as pathfinder. The small twin jet, with state of the art navigation systems, guided the Daggers near the southwest coast of West Falkland, from here they established a northeast heading towards the Falklands Sound, intending later to turn east crossing souther East Falkland and making the attack run on Port Pleasant from the west. The weather had worsened, with low clouds and rain and snow squalls. This made the formation to deviate slightly from the planned route to avoid it. Then while flying across the Falklands Sound on a northeast heading, they had the surprise of their life, at about the ten o'clock position was the frigate HMS Plymouth, the ship was heading to the relative security of San Carlos Water sheltered waters after a night spent in shore bombing duties. It was about 14:00 hrs in the afternoon and the Grupo 6 formation had been detected also by the frigate, the Daggers made a wide turn on a northerly heading to position for the run, the warship had augmented her speed trying to reach the defenses of San Carlos. The Daggers attacked across the port side at an IAS of 550 knots. Rodhe was on the right, Ratti on the center and Gabari by the left, behind in line abreast were cimatti and Martinez and despite the 20mm AA fire and a Sea Cat missile fired to them, continued their run and the Plymouth was hit by 30mm gunfire and five bombs —of eight dropped—. Miraculously for the ship's crew neither of the five Mk.82s exploded, although a fire was declared on the stern as a 30mm round set off a depth charge, the subsequent flooding producing a six-degree list. Only one of the Daggers was lightly damaged by flying shrapnel, and that flown by Ratti could not drop the bombs due to generator failure, after the attack the Grupo 6 fighters turned on a southwesterly heading and during the first leg were pursued by a Sea Harrier CAP without success for the latter. On returning to Rio Grande, the five pilots were acclaimed as heros as at the time the Argentines believed that the frigate had been sunk which was not the case. On June 13 the situation was critical for the Argentine forces on the Falklands,

The second batch of Mirage IIIEAs (I-013 to I-019) were delivered in 1980 and from the outset were capable, in the primary intercept role, to carry the short range Matra R550 Magic IR-homing dogfight missiles, besides the medium range, radar guided Matra R530Es. This capability was retrofitted to the whole fleet after the war. The pictures show Mirage I-017 being readied and taxying for a CAP sortie over the Falklands, note the harsh conditions prevailing in the desolate expanses of Rio Gallegos air base, as well as the A-4B in the background. (FAA)

Eight Mirage IIIEAs were deployed to Comodoro Rivadavia and Rio Gallegos in souther Argentine to provide cover for strike aircraft as well as air defence for this area. The second box of four is seen flying to Comodoro Rivadavia over a cloudy background in late April 1982 (FAA via H.Bosich).

as the British were advancing inexorably towards Port Stanley. In what amounted to the last FAS daylight missions it was intended to launch the maximum number of attack sorties against UK forces ground positions in order to buy time for the Argentine Army, but relatively few FAS aircraft were one hundred percent ready due to technical discrepancies caused by the harsh winter climate with a high humidity factor, battle damage and the compulsory maintenance checks. The staff of Grupo 6 de Caza planned four missions which turned out to be their last of the war. Incidentally Escuadrón Aeromovil Nu. 2 had moved some days earlier to Rio Gallegos. The mission developed as follows:

—Fragmentary Order No. 1317— Was launched at 11:10 hrs formed by three Daggers having the mission to bomb British troop concentrations on the slopes of Mt Langdon. Number three had to abort as he could not retract the landing gear after take-off. The two remaining Daggers, when reaching the target area were confronted with bad weather, rain and heavy cloud cover. Through

Mirage IIIEA I-005 being readied for a mission over the Falklands from Rio Gallegos, note the civilian air terminal in the background. Being one of the initial batch, this Mirage has only the Matra R530 capability, and of course for close combat the 250 rounds provided for the two DEFA 552 30mm cannon. (FAA).

several holes in the cloud blanket, they detected a Sea King helicopter and a pair of Sea Harriers, then the GCi operator advised them that two "blips" were going fast towards them. Thinking they were in danger of being intercepted, the Dagger pair forgot about the attack and made good their escape without getting engaged by the Sea Harriers.

—Fragmentary Order No. 1318— Three Daggers crewed by Captain Dimeglio, Lt Aguirre Faget and 1st Lt Roman. Before take-off from Rio Gallegos, Aguirre had to abandon due to brake failure. Both remaining pilots wandered away from the planned route during their approach to the Falklands; suddenly they caught sight of a British helicopter 45nm south of the Falklands Sound, it was a Lynx HAS.2 (XZ233) from No 815 Squadron and belonging to HMS Cardiff's Flight. Both Daggers attacked knowing that the naval helicopter should have alerted all the AA defenses. Dimeglio and Roman made several gunnery runs but to no avail, the Lynx crew knew their trade and had learned to the last word the training on fighter evasion tactics. With low fuel state, the frustrated Grupo 6 pilots had to return to Rio Gallegos.

—Fragmentary Order No. 1323— Three Daggers with ca sign "Zeus", one of them had to abort for technical reasons, of th two others, one had to turn back when about 60mm from th target due to oscillations above-minus 250/300 RPM on the Ata turbojet and until he landed, the pilot did not make a sigh of relie he had flown more than 450 miles of frozen ocean, with an er gine which at any moment could seize-up, not a situation condu cive to tranquility of mind. The remaining Dagger had to retur some time later as a Sea Harrier CAP could had prevented hir from reaching the target and besides, there wasn't a wingman fo mutual support. —Fragmentary Order No. 1324— Comprised a the others, three Daggers with the call-sign "Vulcano", they ha problems with the weather and a Sea Harrier CAP, having to fo get about the attack as they found that the Sea Harriers wer after them and they were not configured for the air-to-air role. A about 16:42 hrs the last "Vulcano" landed at Rio Gallegos, thi was the last combat sortie for Grupo 6 de Caza, ending its shar in the air war which was not too bright either, with only three ship seriously damaged (Antrim, Ardent and Plymouth) but justifiabl as the Dagger was not the right aircraft for these long range anti shipping sorties —the A-4 Skyhawk in spite of being older, wa better suited to this role thanks to its AAr— capability. First Lieu tenant Jose Luis Gabari Zoco reveals his thoughts about the las days of the war and after:

"....As I was new to the Dagger and Grupo 6, during the heigh of the battle I was on operational training and ferrying replace ment Daggers from Tandil to San Julian and Rio Grande. With 5 hours of Dagger time, on June 1 I reported for duty with Escuadró Aeromovil at Rio Grande. Once there I received several briefing on the war situation, Sea Harrier tactics, bombing profiles an ship recognition. The small Rio Grande base was overcrowded with Navy A-4Qs, Super Etendards, ancillary types from the latte service and our Daggers. Also the runway was shared for civilia use. The cold was intense and frozen mud was everywhere, th aircraft were dispersed in order to minimize the effects of a com mando raid or air attack —a pair of Shafrir armed Daggers wer kept on dawn to dusk readiness with the pilots and ground crew sheltered in a nearby hut— Quarters were clean and warm al though some pilots had to sleep in hotels in town. There was a

During the late eighties, and before the air superiority gray was started to be applied to the Mirage/Dagger fleet, at least one Mirage IIIEA, was painted in this attractive light blue and gray scheme. (author collection)

certain "atmosphere" like the old WWII movies, as the briefing room and the officers mess, there met Air Force, Navy, Army, Coast Guard and civilian aircrews (whose aircraft had been impressed by the FAA), all swapping stories about flying and women (very typical indeed), in spite of high losses, morale was high and most of us thought that the war would continue for a month or perhaps two and so most of us would get killed, also the possibility of conflict with the Chileans was high, as it was clear they had helped the British in some way or other, but we never thought about tomorrow.

I only flew a couple of combat missions, on June 8 when we attacked the frigate Plymouth and on June 13. The next day (June 14) we did not fly as the weather over the mainland and islands was very bad, during the evening we heard about Menendez's (C.O. of Argentine forces in the Falklands) surrender which shocked and demoralized us, although during several days following it, we were on a high readiness state just in case higher authority requested an all-out attack against British forces. That never came and towards the end of June we returned to Tandil, the war was over and we had lost five good friends..."

From May 1, to June 13, Grupo 6 de Caza had flown a total of 133 combat sorties, 88 of which engaging in combat, while the total planned was 160. The Dagger was the mainland-based combat aircraft which suffered more losses, 11 machines, with six pilots ejected and five killed.

During a few days after the surrender of General Menendez and his troops on June 14, Grupo 6 Daggers were kept on a high state of readiness, as narrated by Jose Luis Gabari, but towards the end of June most of the unit returned to Tandil, although a detachment was maintained at Rio Gallegos for a few months. Meanwhile Peru in a rare act of South American solidarity, sold to the FAA ten Mirage 5P for US$ 14 million which covered most of the Dagger losses. They were delivered too late to participate in the conflict, but were active during the uneasy truce which followed after the ceasefire being on standby at Rio Gallegos also. Shortly afterwards, the aircraft were ferried to Rio Cuarto maintenance center in order to make several modifications to bring them

A few months after the war 1st Lt. Cesar Roman poses besides Mirage 5P (C-430) at Tandil air base. (Roman)

to FAA standards, (VHF radio, VOR, ILS, instruments, starting system, etc). The Mirage 5Ps took the serial numbers of lost Daggers (C-403, 404, 407, 409, 410, 419, 428, 430, 433, 436). These Mirages served with Grupo 6 until 1987 when they were delivered to X Brigada Aérea, Grupo 10 de Caza "Martin Fierro" at Rio Gallegos, replacing Mirage IIICJ which were concentrated with Escuadrón 55 (named for the 55 Argentine AF personnel killed during the Falklands War) at El Plumerillo, Mendoza. By the late eighties they were progressively upgraded with Argentine developed avionics in a program known as MARA.

In June 1982 Peru sold Argentina ten Mirage 5P fighter bombers for $14 million. These machines in their original desert camouflage, were pressed into Grupo 6 service replacing Daggers lost and even taking their serials. Later these machines were assigned to the newly activated Grupo 10 de Caza at Rio Gallegos. (author collection)

POSTWAR DAYS

Rio Gallegos air base, October 14, 1982, three months have passed since the conflict with Great Britain for the control of the Falkland Islands ended, I climb down from a Grupo 9 Fokker F-27, it is the southern hemisphere spring, but in spite of this, a chilly wind welcomes me after I reach the ground. As I was a FAA host, the base C.O. was expecting me and after the protocolary salutations together with a welcome late lunch, I went to see the operational area and salute the people there.

Rio Gallegos is a big and desolate air base, sharing their installations with a civilian air terminal, and with few commodities, like the Patagonian region where it is located, one feels there a genuine frontier spirit. The Ford pick-up van with a grinning FAA NCO at the wheel took me over muddy perimeter roads until arriving to the area where the Hardened Aircraft shelters are located. There I see several Daggers and even a pair of ex-Peruvian Mirage 5s. Then I hear a far-off thunder and then a whistle, it sounds very familiar, like Atar turbojets. I raise my head, and there I catch a pair of Daggers on the break and initiating the downwind leg... I meet the Grupo 6 detachment C.O. Major Carlos Napoleon "Napo" Martinez, and we soon started to swap flying stories. He introduced me several other Grupo 6 pilots. talking with them, I felt in their voices and feelings some bitterness and frustration by the surrender and the heavy loss of comrades. I went inside one of the HAS where one of the Daggers rested loaded with bombs on a centerline MER, while the trolley supported part of the bombs weight. They were ready just in case the conflict flared-up again. I looked in detail at the Dagger, which had a white "kill"

Photographed against an impressive backdrop, a Grupo 10 Mirage 5P flies near one of the unforgiving peaks of the Andes mountain range, near the Chilean frontier. (FAA)

marking of a Royal Navy warship over its left intake. In some parts the paint was peeling off, showing the lighter shades of their previous Israeli scheme. Suddenly, as if somebody shouted in my hear, I realized that it was a truly war veteran of the South Atlantic and Middle-East conflicts, she probably had avoided MiGs, SAMs, AA fire and Sea Harriers, perhaps she even killed one or more Soviet-made interceptors and her bombs pierced the thin alloy hulls of a Royal Navy warship, but indeed she had survived and was there ready for another round if the humans needed their services. I touched her cold aluminum skin, and I notice a very strange feeling, as if the Dagger was a living being, wishing that she could tell me about her war adventures under two flags and half a world apart, this makes me sure that some aircraft have a soul within them..."

The words were told to the author by Jesus Romero Briasco, then an Ala 11, Spanish Air Force, Mirage III pilot, who in October 1982 went to Argentina in a short exchange tour with Grupo 8, and at the same time visited several FAA air bases, Rio Gallegos among them.

"RATON" FLIGHT LAST MISSION

21 May, 1982 dawned very cold but clear at San Julian, a civilian airfield located close to the town of the same name in the Patagonian shores, between Comodoro Rivadavia in the north and Río Gallegos in the south. In this provincial airfield, converted in a rudimentary air base, the small parking apron had been augmented by adding PSP (Pierced Steel Planking) in order to provide a minimum dispersal area to the two combat squadrons deployed there since the end of April, an A-4C Skyhawk unit and Escuadrón Aeromovil Num.2.

Guillermo Donadille had arrived to San Julian with the rest of the squadron on April 25 in a non-stop flight from Tandil air base in the province of Buenos Aires. Guillermo was then a Captain in the Fuerza Aérea Argentina, with close to 3.000 flight hours, half of them in fast jets: A-4B/Cs and Daggers and he was a good and competent stick and rudder pilot. On May 1 he flew an escort mission for a strike-configured Dagger flight in the course of which some British warships were lightly damaged.

During the next twenty days he logged many hours in cockpit alerts and a few scrambles, all unproductive and terribly boring. The events of such a dramatic day as was May 21, are described more graphically by this Grupo 6 flight commander:

"""......Yesterday (May 20) I received a parcel from home containing letters, a scarf and some chocolate bars, I found the letters, mainly written by my children, deeply moving.

Today (May 21) I woke up before dawn, then I went out of my room to see the sun rise over the eastern horizon, it was very cold and windy but clear skies promised a good flying day. Soon

During the late eighties the Mirage 5Ps received a limited upgrade with local avionic known as Mara. By May 1995 eight survivors, of the ten delivered in June 1982, were in service with Grupo de Caza "Martin Fierro". Illustrated is on of these parked by the monument to the fallen pilots recently erected at Rio Gallegos. (FAA).

afterwards I started my alert duty as section leader for any possible escort sortie for strike aircraft. As a matter of fact the sleek Daggers are configured for this role: three 1.300-liter drop tanks, two Sahfrir 2 AAMs and 250 HEI rounds for the two DEAF 30mm guns. Just after ingressing into the pilot's room (the airport's hall), we were informed that the British were making an amphibious landing at Isla Soledad (East Falkland) near to the northern mouth of Estrecho San Carlos (Falklands Sound).

We must change the weapons configuration of the Dagger, loading all with bombs. Initially there was a great deal of confusion: orders, counterorders, nerves. Take-off time for the first flight approached and as I was scheduled to participate in the second, I helped as much as possible before departure to my friends from the first wave, consisting of six Daggers. The 6.500-ft runway was a little on the short side with the Daggers at their maximum take-off weight. I prayed for them to return in one piece. With nervous expectation I waited, then almost two hours later the Daggers appeared over the east, all six of them!. After landing and climbing out of the cockpits the pilots, still with the adrenalin flowing, narrated their experiences with the British air defenses, and all at the same time: "very thick", "terrific", were some of the remarks; another pointed: "San Carlos Water is full of ships". Certainly it was a full scale amphibious operation and not a feint as we thought before... The aircraft were checked out and two of them were found to have some rifle-calibre bullet or shrapnel holes.

Now I began to get ready for my mission. I was assigned as leader of a three aircraft flight in number C-403, with Major Gustavo Piuma in C-404 as number two and 1st Lt Jorge Senn in C-407 as number three. This latter pilot had been my pupil at the Aviation School several years ago and I taught him to fly while I was an instructor there. During the exhaustive briefing we got all the necessary information: radio frequencies, type of formation, armament load, weather forecast enroute and over the target, probable Sea Harrier CAPs, etc.

Afterwards we donned the heavy and cumbersome personal equipment, wool inner clothing, pullover, anti-exposure rubberized flight suit, flight boots, anti-g suit, survival and life vest, bonedome, oxygen mask and flight jacket. To all this gear I added an individual touch, a thick scarf with Grupo 6 colors —blue and dark red—. After the external walk around, noticing that the two bombs, and three drop tanks were correctly fixed, I climb into the tight fitting cockpit and with the help of the crew chief I got the harness fastened and all other connections. After wishing good luck with a worrisome expression in his eyes, he removes the ladder and I close the canopy. Then I am isolated from the outside noises and chilly wind, immersed in my world of instruments, switches, dials, which start to come to life through the actions of my gloved hands. Through the earphones in my bonedome, I hear —caused by the mask not being properly strapped— the slightly distorted voice of my number three - "Raton-3": "Ready to start-up". I did not hear number two checking in, time for take-off is nearing and it is necessary to accomplish the schedule; to hell, if he is not ready, I will leave him behind. I give the order to start the engines. Diluted by the noise of the Atars, I hear faintly number two; he has a small technical discrepancy and begs me to wait for him as he does not want to abort the mission. The other flight ("Laucha" also with three Daggers) was ready and I called its leader, Cesar Roman, telling him about our problem and advising that the "Lauchas" should take-off first, as it had been scheduled that we should go first, but that was the only answer in order

Spanish Air Force Mirage III pilot Jesús Romero Briasco (left) photographed during his short exchange tour with Grupo 8 de Caza in October/November 1982, in company with Major Yebra, one of the unit's combat veterans. Major Romero later commanded 111 Escuadrón of the Spanish AF in the period 1985/1987, then equipped with Mirage IIIs, then left the air force and went to fly airliners. (Romero)

A pair of Mirage IIICJ from Grupo 4 de Caza —Escuadrón 55—, overflying the outer Buenos Aires. In total during late 1982 the FAA received 19 Mirage IIICJs and three IIIBJs, all truly combat veterans, many of them aces, from the Chel Ha'avir. (FAA).

to accomplish the established TOT (Time on Target). This change proved fatal to us.

We became airborne on the last feet of San Julian's short runway due to the heavy load carried by our trusty Daggers, it was an agonizing experience. The shore was left behind with the swarms of seagulls. Now we are surrounded by an unforgiving ocean. The emotionless voice of the controller reminds me that I am on course. Also the good weather is left behind, just ahead I can discern heavy cloud build-ups. Near the islands we descend to about ten meters over the dark gray ocean, and at the same time augmenting the speed to 800 km/h. Senn sees the Sebaldes Islands (Jason Islands) to our left, I swivel my head and through the mist and rain showers I can barely see this small group of little islands which serve as a navigational fix. We are about 20 seconds ahead of the estimated time and a little off course.

A few minutes later, we cross the northwest coast of Gran Malvina (West Falkland) Island; the weather gets worse with heavy rain showers and the visibility diminishes alarmingly, I fear for the mountains ahead because we are going very low.

With a glance to both sides, I could see the wingmen at the same level I am flying and balancing up and down in the turbulent air. "Three minutes from the target" I call, "Accelerate now!". I push the throttle lever forward to maximum dry thrust.

We are going very fast over a yellow-greenish undulating landscape, to our sides rise inhospitable hills, while a cloudy sky bathes the area with a gray light. "One and half minutes from the target", my muscles tense instinctively while my mouth gets drier by the moment; as a reflex act I lean forward in the Dagger's small cockpit, getting my attention in the gunsight, through its illuminated signs I can see the ground below, too close for comfort. Following this course, we will go into the sound directly without any need to make corrective maneuvers, thus giving less advantages to the enemy.

"Sea Harriers at three o'clock!!" - the tense voice of Jorge Senn almost caused my heart to stop - I swivel my head and to our right I can see the unmistakable silhouette of a Sea Harrier about 300 meters higher than our flight. I have the strange feeling that his wingman is behind us. Almost at the same time the British pilot saw us and started and started a diving turn towards us.

"Eject bombs and tanks, break right, go!!" I order, at the same time as I do the same in order to engage the Sea Harrier head on. One of my wingman hesitates, I repeat the order. Now I see his underwing stores separate from his Dagger, which without so much drag leaps like a frightened rabbit crossing in front of me.

The British pilot maintains a convergent heading to my aircraft in a shallow dive. I press the gun trigger from a distance of 700 meters, I think that the flashes from the guns surprised him steeping the dive angle, my burst goes high over his cockpit without doing any harm.

I make a 90-degree bank and initiate a rudder dive, lowering the nose and trying to avoid losing the Sea Harrier from sight underneath. I fire again expecting that the Sea Harrier could "catch" any 30mm round. Now the ground approaches very fast indeed, and my windscreen is completely filled by a dark gray fuselage flanked by two big intakes and a pair of short swept wings.

Sick to the stomach, care with the yawl, I feel the g-forces pressing me against the seat, the anti-g suit inflates preventing the black out. I see one of my wingman passing over my left side on a level hard turn. I half roll the Dagger and while upside down I see this aircraft going away with a bright flame pouring from the tailpipe, indicating he is in full afterburner.

Then I hear a small detonation, not very loud, as if somebody crushed a paper bag besides my ear. The my aircraft gets out of control, first pointing into the clouds above, then starts a frightening pitching motion flattening me against the seat or making me to float in company with all the dirt rising from the cockpit's floor.

Suddenly, the Dagger starts a series of fast rolls —parallel to the ground— the control stick feels soft, without any command capability. Due to the speed and closeness to the ground I thought that my life on the ground was reaching its end, at the same time a strange kind of fatigue invaded my body, but then the surviving instinct which is so deeply rooted in combat aviators, reacted and with both hands I pull the lower ejection handle. Again God pro-

Close-up of a Escuadrón 55 Mirage IIICJ. These Mirages lacked the Cyrano Ibis fire control radar and were provided with Atar 9C turbojets, most of them serving with the above unit, part of Grupo 4 at Mendoza. (author collection)

'he Mirage IIICJs were serialed as C-701 to C-719, while the IIIBJs were serialed C-720 to C-722. The machine photographed, seen while in open air naintenance, was painted in a darker earth scheme, while assigned to Grupo 10 de Caza at Rio Gallegos, note its insignia applied on the tailfin; eventually ill were concentrated within Grupo 4 at Mendoza. (author collection)

ected me as I eject when the Dagger is almost upside, the chute pens and seconds later I make a very hard landing, I can not ee very well as my eyes are affected by the high speed wind lasting at my face during the ejection, but thanks to God I have o other injury. I conceal the parachute and get away from the lace while hearing the 30mm rounds cocking-off by the fire as he unfortunate C-403 crashed about 300 meters from where I anded.

I walk for one hour and fifteen minutes following a line of elegraph posts. I find an old plough and a wooden hut in ruins, reaking a door, I take a pair of long boards, assembling a small helter as the night approaches. I fill a sackcloth with dirt as some hort of makeshift pillow and I am ready to spend the longest ight of my life, my family and my Grupo 6 comrades back at San ulian and Rio Grande seem very, very far away in the cold night, n incredible cold which seems to come by waves, and which revents me from sleeping during those endless hours.

With the dawn and as the sky becomes clearer, my worry bout any possibility of having any kind of back injury due to the jection waned, as this kind of problem is common and appears everal hours after an ejection. Fortunately I have no trouble getng to my feet.

That day (May 22) I walk about 25 kilometers following the dications of the small compass carried with the survival gear, esides, I know a bit about the Islands geographic marks, thanks o having carefully studied the maps, I arrive without any deviaons in Port Howard at about 15,00 hrs where there is an Argenne Army Infantry Regiment quartered.

More dead than alive with tiredness, and with the beginnings f some dehydration, but happy knowing that I should be able to end messages to my family and my comrades at San Julian, I tayed with the army, then after several adventures with other ejected pilots —including my wingmen— we flew by helicopter to Port Stanley and towards the end of May we returned to the mainland aboard a C-130....." This first hand account provides details of the drama which involved most of the Fuerza Aérea combat sorties from the mainland. They were very courageous pilots, but were not very well versed in the arts of modern air-to-air or air-to-ground tactics, anyway, they provided ample proof of their valor.

THE OTHER SIDE OF THE COIN

If in the above lines a Dagger pilot tells his dramatic story, if could be very enlightening for the readers that the pilot who shot him down, Lieutenant Steve Thomas, RN, from No 801 Squadron makes his account of this particular air combat:

"....On the third sortie of the day (May 21), I was flying Sea Harrier 009/ZA190 and was wingman to LtCdr Nigel Ward patrolling at 1.000 ft in valleys 15 miles northwest of Port Howard. I saw two Daggers together flying east fast at very low level. I maneuvered in behind and locked and fired a missile at nearer aircraft. The Dagger blew-up and crashed, leader of element pulled-up and right (presumably going home fast!) I locked and fired at him my second and last AIM-9L and the missile guided well and went over his port wingroot causing a bright orange flash against the aircraft. We could not see what happened to him as Ward (and I) had seen a third Dagger crossing ahead of us, which Ward destroyed with a Sidewinder —total time about one minute—. It was subsequently found that my second Dagger was lost also, fortunately all three pilots ejected safely..."

Modern air combat is a world apart from other forms of air warfare and the above narratives of an Argentine pilot and his British colleague, the victor, provide ample proof of that assertion.

Actually most of the Grupo 6 Daggers have acquired this attractive gray scheme, as can be seen on this machine which also has applied in a darker gray the new Grupo 6 insignia a eagle's head, note the Mk.82s and Shafrir IV AAMs. (Padin)

MIRAGE III INTERCEPTORS

During the late sixties the FAA staff was aware about the inadequacy of the country's air defence, which was vested on the then obsolete F-f Sabre and the still older Gloater Meteor. A study group was formed and after several evaluations, the F-e was selected as the new fighter/interceptor, but the US Department of State vetoed the sale and the FAA had no alternative but to acquire the second best in their list, which was no other than the Mirage IIIE.

The program contemplated first was to build under license about one hundred examples, which was later reduced to fifty and finally, mainly due to the Argentine economic difficulties, a contract was signed with Dassault for the acquisition of ten Mirage IIIEA interceptors (I-003 to I-012), a simplified variant of the E without the doppler radar and other systems but with the addition of equipment specified by the FAA. Also two dual-seat trainers were acquired as Mirage IIIDAs (I-001/I-002). A group of highly experienced pilots were sent to Dijo, France for conversion to the delta fighter with ECT 2/2, which after their return acted as instructors. Deliveries started in September 1972, transported to Argentina in the cargo bays of FAA C-130 Hercules. In July 1973 the first flights were undertaken and in that month the then known I Escuadrón de Caza Interceptora (1st Fighter-Interceptor Squadron) was activated, based at Mariano Moreno air base, some 25 km south of the city of Buenos Aires. In order to provide the unit with a more formal context and to follow the typical designation system used in the FAA, on 9 December 1975, VIII Brigada Aérea was activated and soon afterwards Grupo 8 de Caza. The Mirage IIIEA main role was the air defence of the republic's capital. Since the first aircraft were delivered in, until the Falklands conflict in April 1982, two Mirages had been lost: I-009 in 1976 on a flat spin, the pilot 1st Lt Garcia ejected safely, the second lost was I-001 in March 1979 also on another unrecoverable flat spin, both pilots, Captain Huck and Comodoro (Colonel) Viola ejected. To augment the fleet and to cover the attrition, seven new Mirage IIIEAs were acquired in 1979 (I-013 to I-019) with deliveries completed in 1980, being activated within Grupo 8 a second squadron. Since the aircraft entered service in their primary role of interceptors —their secondary tasking was that of fighter bombers— they had, besides the two DEAF 30mm cannon, the MATRA R-530 medium range air-to-air missiles, both in the infrared guided and semi-active radar guided variants. In 1980 a batch of MATRA R-550 Magic short range

IR-guided AAMs was acquired starting to arrive to Grupo early in 1982.

During the Falklands War the unit lost two Mirage IIIEAs (see later, and in December 1982 two new dual seaters (I-020/I021) were received from France, thus complementing the Mirage flight simulator for training new pilots to the type. On 17 March 1988 the two Mirage squadrons were officially transferred to Grupo de Caza as Escuadrón III and Escuadrón IV, but later amalgamating as Escuadrón I, after Grupo 8 and its respective Air Brigade were deactivated.

Originally the Mirages were received in a scheme similar to the USAF's SEA, but during the late eighties the whole fleet has been painted in air-superiority grey. All maintenance of the airframes, engines and avionics is performed in country by the Rio Cuarto rework facility, including damage repair. During the last years the Mirages have received a few small upgrades, mainly to improve the capabilities of their RWR equipment, but a decision will be taken in a near future whether to make on them an extensive upgrade, including fitting a new radar replacing the mediocre Cyrano IIbis, or acquiring second-hand ex-USAF Lockheed F-16A/Bs to supplant entirely the Mirage IIIEA/DA fleet towards the end of this decade.

On 2 April 1982, Grupo 8 de Caza formed part of Comando Aéreo de Defensa (Air Defence command). With the FAA mobilization on that day, the unit was put on a high state of readiness this was soon relaxed. But as the days passed and the FAA made to the idea that the British were not joking and that there existed also a possibility of Vulcan attacks, the Mirages were again alerted for using on their main role of interceptors/all weather fighters and not as fighter-bombers. The maintenance department worked overtime for readying the maximum number of operational aircraft, and of the 16 EAs available, twelve were fully operationally fit for combat. Meanwhile the pilots were instructed in the firing parameters of the new (for them) Matra Magic IR AAM, also studying all available information about the Harrier.

The insignia of the today disbanded Grupo 8 de Caza, the olive branch within the Mirage signifies "combat proven". (Padin)

Mirage IIIDA I-001 just before delivery from Dassault assembly plant at Bordeaux, it was lost in March 1979. Late in 1982 another 2 two-seaters were delivered from France (I-020 and I-021). (Dassault)

Also the possibility was studied of basing the Mirage IIIs at Port Stanley airfield, but this option proved to be impossible as the runway was too short, 4,100 ft, for operating with a minimum of safety with external loads. After the Fuerza Aérea del Sur (Southern Air Command) was activated for coordination FAA operations in the theatre, Grupo 8 de Caza deployed eight Mirage IIIEAs to Comodoro Rivadavia air base in the province of Santa Cruz, a few days later, four were sent to Rio Gallegos in the same province but located more southerly. Thus four remained at Dr Mariano Moreno air base as reserves and for air defence of the capital. The eight sent to the south received the designation "Escuadrón Caza Interceptor" being under the command of Comodoro (Colonel) Carlos Corino, Grupo 8 C.O. For the air defence of the Patagonia and Tierra del Fuego bases and military installations, the FAA had Westinghouse AN/TPS-43E radars at Comodoro Rivadavia, Rio Gallegos and Rio Grande.

The worst handicap for the Mirage III was the more than 400 nautical miles separating Comodoro Rivadavia and Rio Gallegos from the Falklands. The Daggers with their extra 500-liter upper tank aft of the cockpit could remain on station for about five minutes, meanwhile the Mirages would reach the vertical of the islands at great height without having the possibility of going to lower altitude to engage the Sea Harriers due to fuel considerations, thus before the start of hostilities, the Mirage III had already lost the battle for gaining air superiority.

On April 30 the events seemed to happen at once: the British were near the islands and the mainland-based Mirages were on instant readiness, in spite of the intense cold, while the radars probed continuously in search of hostile contacts.

As a consequence of the 04:45 hrs raid by an RAF Vulcan bomber to Porp Stanley airfield (BAM Malvinas for the Argentines on May 1, the FAS issued Fragmentary Order number 1090 in which two Mirage IIIEAs with the call-sign "Fiera" were launched from Rio Gallegos, with Major Sanchez as section leader. The mission purpose was to provide top cover to Port Stanley airfield just in case another Vulcan raid developed. At 07:30 hrs and after having been a few minutes on station, the pair returned to Rio Gallegos uneventfully. This mission and subsequent ones were controlled by the AN/TPS-43E radar located near Port Stanley and a Command Post installed in town, known as the CIC by the Argentines.

The tearing sound of two Atar 9Cs in maximum afterburner power thundered over the desolate expenses of Rio Gallegos, announcing the take-off of "Tablon" section. Two minutes later "Limon" section was launched flown by Captain Garcia-Vuerca and 1st Lt Perona. When at 20,000 ft over the Falklands they were vectored to cut across an A-4B flight and a Sea Harrier CAP which was trying to bounce the former, the Royal Navy fighters broke the engagement. Seconds later "Limon" section received vectors to intercept another Sea Harrier CAP, this was from No 801 Squadron and flown by LtCdr Ward and Lt Watson. Garcia Cuerva and Perona pursued the Sea Harriers until being 15 miles behind them, then the Sea Harriers made a quick 180-degree turn, engaging head-on. The distance was now shortening dramatically, then the British pilots saw as a Mirage (which were higher) fired one or more missiles, which due to the excessive distance, fell short. In fact only a Matra R 530 was fired and by mistake without a full radar lock-on. The other "launchings" were the 1.700-liter drop tanks which when ejected for better dogfight capabilities, produced a trail of vaporized fuel, easily confused with a missile from distance. Both Mirages were critically short of fuel and they had to break the engagement and return to Rio Gallegos, on which they landed with very few liters to spare, while before that, "Tablon" section had returned uneventfully.

The next Grupo 8 mission was again CAPping over the Falklands and to this effect, "Dardo" section was launched at 15:30 hrs and "Bawdry" did it at 15:53 hrs. "The "Betrays" were again Garcia-Cuerva and Perona, engaging north of Pebble Island at 16:10 hrs with a 801 Squadron Sea Harrier CAP flown by Flt Lt Baron and Lt Thomas, the scene for the first air combat of the war was set. The Grupo 8 pilots approached with an altitude advantage and head-on trying to entice the Sea Harriers in a vertical scissors maneuver. During this approach, they only caught sight of thomas, as he and his leader were in the standard widely separated line abreast defensive battle formation, and the Mirage jocks not perceived as Baron made a pincer movement towards the northeast, approaching from the rear side. The Argentine fighters had been visually detected by Baron at about eight miles, when the distance was reduced to almost half, Thomas saw what seemed the launching of missiles against them, again the jettisoning of drop tanks had been confused. Perona and Thomas crossed very close, the Argentine pilot could see with all detail the Sea Harrier lower surfaces painted in dark gray as well as the drop tanks and Sidewinders. They were at 12,000 ft. He tried to follow visually the Sea Harrier and when initiating a left, his Mirage I-015 was hit by an AIM-9L fired by Baron which was aboard Sea Harrier XZ453, while the two Argentine interceptors were in a smooth (about 2.5 g's) turn, while the British pilot had made a hard (about 6 g's) turn to position behind, getting a lock-on with his left hand Sidewinder at one mile approximately. It was the Sea Harrier first kill. For Perona the missile hit came as something of a surprise, fortunately he could eject from his crippled

Mirage IIIEA I-003 seen at Bordeaux prior to an acceptance flight besides on the prototypes of the Mirage IV nuclear bomber. (Dassault)

fighter, and now follows the narrative of those events as told by him:

"....The tactical idea at that moment was that each of us lock into a one versus one combat. At that moment our aircraft and a Sea Harrier crossed each other, so I started what is known as a scissors maneuver, consisting in raising the aircraft to try to gain as much height as possible, turning and banking to port and starboard in order to try to gain advantage over the enemy, which is flying roughly parallel to you, and placing one's fighter at his six o'clock position, but the Sea Harrier characteristics could make it faster to slow down in the air, besides I had trouble in controlling my Mirage at slow speed, due to a drop tank hang-up, so when I took a glance to one of the sides, I noted that we were separated by about 100/200 meters, I followed it until disappearing from my field of view through the top aft side of my cockpit canopy, just immediately when I was starting to maneuver to eyeball the Sea Harrier again, I noticed a big jolt, my Mirage started to shudder

(Above and Opposite) In 1988 Grupo 8 de Caza was disbanded and its assets were sent to Tandil air base to augment Grupo 6 de Caza, which in 1995 had two squadrons of Daggers and one squadron of Mirages. (Padin)

badly and I lost control, while the warning lights of the panel illuminated in quick succession. After futile attempts to control my machine and seeing they were of no use, I ejected from the crippled plane. I succeeded in heading my parachute towards land, as to fall into the water on these latitudes is a dramatic contingency, but when I was some meters above the ground the wind pushed my parachute towards some rocky terrain, making contact with the ground in an unorthodox manner, suffering wounds to both legs..."

Perona was rescued some time later and once in Port Stanley evacuated to Comodoro Rivadavia in a C-130 Hercules, spending most of the war in a military hospital recovering from his wounds. Moments later Thomas in XZ452 approached to firing range from the other Mirage (I-019) flown by Garcia-Cuerva, launching an AIM-9L, but who better than this Sea Harrier pilot to explain the following action:

"...I got a visual at about eight miles and saw that it was a Mirage. Also I could see his wingman, just to the right and behind, I could not believe that a pair of fighters would come in along like that. Their formation was poor, what the Americans call "welded wing". I was trying to lock a Sidewinder onto him but the missile would not acquire as the solution was wrong. At five miles the leader launched a missile, but I saw it diverge and go down to my left at the same time something came off the second aircraft, it was tumbling. I began turning hard to the right, and passed about 100 ft above the top of their leader. I could make out every detail of the aircraft, its camouflage pattern, and see the pilot in his cockpit. I continued my turn towards the wingman who was about 500 yards in trail on the leader's left. As I was in the turn, I passed over the top of Paul, as he was moving in to attack. I saw his guns firing. Then the rear man rolled off the bank and started a gentle climb, seemingly unaware that paul was sitting behind him. My opinion was that these pilots were not very good, tactically they did not seem to know what was going. They both seemed aware that I was there, but I do not think they knew what Paul was up to.

I saw the Sidewinder launched by Paul streak towards the wingman, it exploded and the rear end of the Mirage became a bright yellow torch. The front half of the aircraft remained more or less intact and shedding pieces, continued on and upwards in a ballistic trajectory.

Meanwhile I was continuing my turn and the enemy leader was doing quite a descending turn to the left, going down very fast towards the top of the solid cloud cover at 4,000 ft. I rolled into a vertical descent behind him locked one of my missiles and fired it. The "Lima" streaked after him and just before he reached cloud I saw it pass close to his tail, then both, the aircraft and the missile vanished...""

It seems that the Sidewinder exploded close to the low aft section of the Mirage, some shrapnel puncturing the fuel tanks and damaging the control runs. In an effort to save the precious aircraft Gracia-Cuerva tried to land at Stanley airfield. After about five minutes, he reached near the runway. The controller ordered Garcia-Cuerva to eject, but the Grupo 8 pilot was trying by all

Argentine Mirage III pilot in the typical flying kit, before an air-to-air gunnery sortie, the small poster on the noses means that the aircraft's guns are loaded. (Padin)

means to save the damaged interceptor despite the risks of a damaged runway, as that of Port Stanley was. The AA defenses were advises about the approach of a friendly, then on short finals, he jettisoned his Magic missiles on inert mode —pulling the emergency handle in the cockpit— the nervous crew of an Army 35mm gun, thought it was a British aircraft attacking and opened fire, hitting almost immediately the Mirage, Garcia-Cueva last words were: ""Hell, they hit me.."" the aircraft, enveloped in flames, rolled inverted and crashed into the water to the south of Freycinet Peninsula, he was the only Grupo 8 pilot killed during the conflict. Soon after, "Dardo" section returned to Rio Gallegos without engaging the Sea Harriers.

Thus finished, after twelve sorties over the Falklands, the first day of the war for Grupo 8 de Caza. During those missions the Mirages of the first batch carried a configuration of two 1.700-liter drop tanks and an R 530 radar guided AAM, while the second batch added the two Magics. This later configuration was discouraged in the pilot's operating procedures manual, but the advice was not considered and the aircraft thus loaded operated with the usual margin of safety.

This first day of combat shocked the Mirage pilots, besides losing one of their comrades —an experienced flight commander— they had lost two almost brand-new Mirage IIIEAs and several negative aspects came to the surface which had to be considered during the remainder of the conflict. The first was the clear disadvantage in the air-to-air area with the duo Sea Harrier/AIM-9L, this missile was more advanced than the R 530 or R 550 carried by the Mirages. Another matter was the use of not very developed-aggressive tactics in air combat. In second place came the lack of range which made it impossible to reach the operations area, dive to lower levels to engage the Sea Harriers and return. For all these reasons in the evening of May 1, the feelings among Grupo 8 de Caza pilots were of frustration, to be bettered by an enemy (the Sea Harrier) which they had underestimated, mainly the combat readiness of its pilots. Consequently there was no point risking the few available Mirage IIIs interceptors in engagement in which they did not have a clear cut advantage. Besides, the FAS Staff was logically worried that the Vulcan's next raids could be the southern Argentina air bases, mainly Rio Grande or Comodoro Rivadavia, accordingly, Grupo 8 with its 24-hour intercept readiness mission at Rio Gallegos and Comodoro Rivadavia, was ready in case any threat was detected by the EW radars. These alert duties which were maintained during all the conflict and even a few weeks afterwards, were terribly boring, inside the Mirage III narrow cockpit, enduring very low temperatures, without really knowing for what kind of missions they would scrambled, without knowing if after these painful expectations their services would be required. This is the main reason why fellow airmen from other FAA units called them the "trench foot pilots". In the following lines, Major Sanchez, Grupo 8 Operations Officer during the conflict, relates some of his experiences while on alert duty at Rio Gallegos:

"""....I made a couple of alert scrambles, searching for low speed contacts, probably helicopters flying very close to the Argentine/Chilean border line in Tierra del Fuego, usually the bogey entered Argentine territory, we scrambled, then returned to Chilean territory, the radar lost it and we remained on patrol until reaching "bingo" fuel. On another night we patrolled over Tierra del Fuego, near Rio Grande looking for some bogies which we could not fix. Although we did not have at the time RWR (Radar Warning Receivers) on the Mirages, through the IFF we could identify if the Chilean radar at Punta Arenas "painted" us. We were on permanent readiness and with the capability to cover any type of mission, as this was a role which we knew perfectly, thanks to the operational exercises which we did regularly. Consequently we had a good capability from the point of view of air defence and interception in all kinds of weather.

I'll like to emphasize that the personnel was trained to endure intense efforts, during some time there remained a pilot per aircraft ration at Rio Gallegos, in spite of this, all the Mirages were on 24-hour alert during more than a week. This alert covered all the FAS bases and also Rio Grande, while at the same time making top cover and decoy missions over the Malvinas (Falklands). When returning from these later missions we had to turn around quickly the Mirages, then standby alert and perhaps scramble again. In one instance, I returned from an escort sortie which lasted for more than two hours, after landing, my aircraft was being refuelled at the ramp, then I received an urgent

An air superiority gray Mirage IIIEA with some of its pilots posing besides it, note the new Grupo 6 insignia. (Padin)

scramble order, so the refuelling was cut short and I launched with only internal fuel...""""

The period of time between May 2 and 20 was of relative quietness, although under the stress of the unknown at to what course the events could take. The British landings in San Carlos and adjacent areas caught the FAS by surprise, as they were not expected there in spite of the clear advice which was the SAS (Special Air Service) raid on Pebble Island airstrip a week earlier. Notwithstanding, the strike missions commenced almost immediately as well as the air cover missions flown by the Mirage IIIs. Consequently in what was later called the Battle of San Carlos between May 21 and 27, the Mirage IIIs made 18 sorties from Rio Gallegos (now all were concentrated on this base). During these six days, the Grupo 8 interceptors had a high availability rate — close to one hundred percent— and all aircraft dispatched arrived to the operations area without engaging the Royal Navy CAP as the Sea Harriers had their most profitable trade at low level and were not interested in climbing to high altitude to engage the Mirages, which due to the range problems explained before could not come to lower levels to engage. Now former Grupo 8 Commanding Officer Carlos E. Corino explain these tactics:

""""....Basically, our aircraft operated only at high level as they had the advantage on these conditions, losing it when going lower to engage. We had plain conscience of our advantages and disadvantages and naturally, the British were also perfectly aware of our troubles. Nevertheless to exploit in a convincing manner our advantages, we should had a better guidance (GCI) capability, and although the Mirage III has a search and fire control radar, guidance had to be provided from the ground based GCI station until being positioned in a favorable attitude on respect of the enemy, that is, to be placed within the range of the onboard Cyrano radar, with a theoretical 25-nm search capability. Our interceptors operated in the extreme edge of their radius of action, and in the better of times could not stay on CAP over the islands for more than a few minutes at high level. The Sea Harrier pilots had a far better guidance capability with a respectable quantity of destroyers/frigates with highly experienced fighter controllers aboard. After the May 1 combat, there was only a few skirmishes without consequences, when the Mirages were on a favorable position, the Sea Harrier disengaged and vice versa if the Mirages lost the advantage in altitude..""""

During the conflict's last period, Grupo 8 made a total of 15 air cover sorties over the Falklands, again without engaging in combat with the Sea Harriers. It is worth to point out that the four cover/decoy sorties made on June 8, in some way carried away the attention of the Sea Harrier CAPs allowing the first Skyhawk and Dagger waves a relatively undisturbed run to their targets. If there was a dramatic escort mission, that took place on the night of June 13 in what amounted to be the FAA's last combat mission of the war and for the drama involved it is interesting to describe it in detail.

About 19:00 hrs on Sunday, June 13, the hated black telephone rang at the Rio Gallegos air base Command Post. A pair of Frag Orders (1326 and 1327) were transmitted to prepare a strike sortie by a section of Canberras of Grupo 2 de Bombardeo to attack British positions on East Falkland and to prepare an escort section from Grupo 8 de Caza. Both Twin jet bombers had arrived earlier from Telew naval air station in order to have a shorter trip to the target. The Mission planning was very complicated for the escort fighters due to range considerations.

It was necessary to compute routes, rendezvous points and schedules exactly so that the Mirages could protect the bombers from the Sea Harriers during the approach and attack phase of the mission. all the crews were at the briefing to establish all the procedures to be followed. Soon afterwards the Canberra section under the call sign "Baco" was launched. Number one was crewed by Captain Pastran (pilot) and Captain Casado (navigator/bombardier) and Number two by Lt Rovollier (pilot) and 1st Lt Anino (navigator/bombardier).

Fourteen minutes later the Mirage III section was launched with the call-sign "Pluton", the leader was Major Sanchez and the wingman Captain Gonzalez. It was an extremely dark night without moon, at 5.000 feet there was a dense cloud cover which extended almost the whole distance to the islands. The two Mirage IIIs, once at their established cruise altitude, initiated the crossing to the Falklands with the navigation lights switched off and in complete radio silence. Due to the night's darkness, both interceptors were using a radar trail formation with the wingman keeping station with the leader using the Cyrano IIbis set.

"Pluton-1" as well as his No 2 were very experienced combat pilots, and it was no less due to the difficulties involved in this mission, the long night crossing over the vast expanses of the ocean without external references or visibility, the deep sense of loneliness, the monotonous engine's droning, in which the slightest drone change in noise level or fluctuating RPM indicator scared the pilot's soul. The radio in complete silence, the outside world darkness, the softly lit instruments and radar display, could create on less experienced pilots a feeling of vertigo, in which the aviator distrusts the indications of his flight instruments at the same time that his senses dictate him wrong impulses. They were flying at 30,000 ft and close to the southern coast of West Falkland, the cloud cover was broken in several places, in fact through one of the holes they noted some brilliant flashes of variable intensity, a kind of storm, but only in one part of East Falkland on a very regular rhythm.

The new Grupo 6 de Caza insignia, the starts are the six pilots from the unit killed during the Falklands War. (Padin)

Line-up of Grupo 4 Mirage IIICJs while on a weapons meet at Tandil in 1988. (author collection)

The Mirage IIIs detected on their Cyrano radars the Canberra section, which were at 40,000 ft in front of them and near the target. Breaking radio silence, they established contact with the Falklands radar, and the controller confirmed that the flashes were artillery shells from British field and naval guns bombing Argentine positions around Port Stanley, although the shelling had not affected so far the radar's operational level. No Sea Harrier CAPs were in the area so the Mirage drivers eased their search.

The Argentine radar guided the Canberras toward their objective, soon the first passed over the IP (Initial Point), closely followed by the second, the bombs fell over the signalled coordinates: British troop concentrations an HQ area near Mt Kent. From the relative comfort inside the fighter cockpit, Major Sanchez could perfectly discern the brilliant flashes caused by the exploding bombs, through some holes in the clouds.

Nearby he noted a missile being launched, which probably was defective as it started to make wild turns. It did not happen the same with subsequent launches, the British had fired four or five of them at the same time, climbing almost vertically in a steady way. Major Sanchez called to Captain Gonzalez through the VHF radio warning him of the danger, although apparently the missiles were locked into the Canberra section, probably "Baco-2. But Lt Rivollier, warned by radio of the situation, launched several flares and chaff bundles, initiating high-g evasive turns managing to avoid the missiles which were lost into the dark night.

The observation of these events distracted the attention of Sanchez and then with his peripheral vision saw a new volley of missiles, these being at twelve o'clock, much lower but climbing very fast, distinguishing perfectly the bright red exhausts, now they were heading towards their aircraft. One of the missiles in particular, grew bigger, seemingly locked on the Mirage. At that moment he was at 30,000 ft and the speed was Mach 0.92, doing a wider turn with the nose pointing to a heading of 045 degrees, the missile was inside the turn but somewhat lower. Reducing engine power, the turn was tightened while at the same time commencing a deep spiral descent. Major Sanchez eyes were glued on the rapid closing light, which was now at the center of an imaginary circle drawn with the Mirage.

He noted that the flashes coming from Mt Kent and Port Stanley had finished, so sky and earth were confused in a frightening black velvet. It was very difficult to discern which way was up or down, being imperative to maintain the eyes glued on the surrounding airspace forgetting about flying by the instruments. Sanchez soon noted the first vertigo assaults, the descending spiral continued, but then the missile exploded in an extremely bright fireball while red-hot shrapnel from the warhead pierced the night sky tracing many red lines. The Argentine pilot sighed with relief, the Mirage had not been hit, but again, fear came to him. A new and bigger missile —probably a Sea Dart—, was seen climbing at a great rate of knots, fortunately for him, it passed by his starboard side still ascending without exploding.

Temporarily out of danger, a quick glance was directed to the instruments, the pilot noting that the descent had brought him to a little below 15.000 ft and to the west of the planned course. Establishing contact with the FAA radar at Port Stanley, the controller replied that he had both Mirages on the screen, but they were almost overflying Goose Green, with Captain Gonzalez flying much lower, almost within the reach of AA artillery and Blowpipe or Stinger missiles. The Mirage section recovered its cruise height soon, at the same time trying to establish radio contact with the Canberras. The only reply came from "Baco-2", which informed that himself had been trying to contact with his leader without success. A request was made to the radar controller, and the feared reply came, any type of contact with "Baco-1" had been lost. This Canberra (B-108) had been downed by a Sea Dart missile launched by the destroyer HMS Exeter, its pilot, Captain Pastran was able to eject, being captured by British troops, while the navigator, Captain Casado fell to his death in the spinning Canberra which crashed into Mt Kent. The three remaining aircraft returned to the mainland, landing at Rio Grande at 23:55 hrs.

With this mission finished the pat played by Grupo 8 in the Falklands conflict, having made a total of 45 air cover and decoy missions over the islands and another 46 over the mainland and territorial waters (alert scrambles and patrols), losing two aircraft and a pilot in combat.

Dagger flight line, in the foreground is a two-seater. (author collection)

The second Mirage nearest the photographer is a Mirage IIIBJ. (author collection)

Having explained before what were the problems of Grupo 8 for not having disputed the air superiority to the Sea Harrier — these apparently poor results were criticized in some FAA circles— , the Mirage III were in fact the authentic reserve for the Daggers and Skyhawks had the conflict lasted longer. Besides their important contribution to the air defence role can not be diminished. After the war, Grupo 8 returned to the peacetime routine and due to the galloping economic crisis in Argentina, drastic cuts were implemented in the air force and the number of flying hours allotted to the pilots diminished constantly, reaching late in 1984 a low figure of five per pilot per month. Detachments were and are made periodically to Rio Gallegos to exercise and provide supplementary air defence assets. As related earlier, in 1988, the Mirages were relocated to Grupo 6 de Caza, but continued with the same taskings.

MIRAGE IIICJ/BJs FROM ISRAEL

To replace most of its war losses, besides the ten Mirage 5Ps and two Mirage IIIDAs received in 1982, the FAA acquired 19 Mirage IIICJ (C-701 to C-719) and three Mirage IIIBJs (C-720 to C-722) from surplus IDAF stocks by US$ 70 million. These aircraft arrived to Argentina between December 1982 and February 1983, having been reengined during the early seventies in Israel with the locally built Atar 9C. They were sent to Rio Cuarto overhaul and repair center for conversion to FAA standards and a complete overhaul. The technicians there found them in good condition despite their age —the oldest had some twenty years of heavy usage—. All had new spares and re-skinned wings during their upgrade periods at IAI, as well as the troublesome Thomson-CSF Cyrano Ibis radar removed and its place fitted a lightweight Elta ranging radar. During late 1983, the first Mirage CJ/BJs were delivered to Grupo 4 de Caza of the IV Brigada Aérea, at El Plumerillo, Mendoza, who turned their seven surviving A-4Cs to augment the A-4Bs of Grupo 5 de Caza of the V Brigada Aérea. The Mirages were and are operated by the especially commissioned Escuadrón 55 (named for the 55 FAA personnel killed during the Falklands War) On 19 March 1984 X Brigada Aérea was activated at Rio Gallegos, with Grupo 10 de Caza, thus creating a permanent fighter presence at this Patagonian air base. Five Mirage IIICJ were delivered on that year to the unit, but were replaced by the Mirage 5Ps (upgraded in the Mara program) while Mirage IIICJ/BJs were concentrated with Escuadrón 55 at El Plumerillo, Mendoza, which also acted as some sort of Mirage OCU.

7

Deltas Over Pakistan

After the 1965 war with India, Pakistan was facing an embargo on arms by the United States which forced the Pakistan Air Force (PAF) to look elsewhere to modernize its combat assets and to narrow the quantitative and qualitative edge that the Indians had with massive infusions of Soviet delivered and license-built MiG-21 Fishbeds. The first order in 1966 was for 18 Mirage IIIEP all weather fighters, with MATRA R 530 medium range missiles, while in place of the short range Magic, AIM-9 Sidewinders were adopted. These aircraft were delivered between 25 October 1967 and 30 April 1969 (see box # 1 for details), being delivered by air by PAF pilots. Followed by three Mirage IIIRPs (reconnaissance fighters) between 10 April and 23 June 1969 and three IIIDP trainers starting on 30 January 1969. In 1970 Pakistan ordered 28 Mirage 5PAs and two additional IIIDPs, and in 1975 the small number of reconnaissance deltas was augmented by ten additional Mirage IIIRPs.

The fourth contract included two Mirage 5DPA2 trainers delivered from 20 September 1981 and 30 Mirage 5PA, of which the last arrived in Pakistan on 29 December 1982. These comprised two subvariants, 18 5PA2 fitted with the Cyrano IV radar (like the Mirage F1) and 12 5PA3 with the Agave radar (like the Super Etendard) in order to provide this variant with an anti-shipping capability with the AM 39 Exocet missile. Later, in 1990, the PAF acquired 50 Mirage IIIs phased out by the Royal Australian Air Force, including 43 single-seat IIIO(A)s and seven two-seat IIIDs. The ex-Australian Mirages are being upgraded and overhauled at the PAF's own Mirage Rebuild Factory at Kamra. Not all of them will enter into operational service, as some have been broken down for spares. The recently refurbished machines have been issued to No. 5 Squadron to top-up its strength which had been reduced to some ten Mirage IIIEPs and eight RP/RP2s after many years of continuous operations.

Between 25 October 1967 and 30 April 1969 a total of 18 Mirage IIIEPs were delivered to the Pakistan Air Force, these were used to reequip number 5 Squadron, which were blooded during the December 1971 War with India, in which they shot down ten IAF aircraft without loss to themselves, also they undertook many ground attack missions, illustrated is Mirage IIIEP number 112 dropping its load of GP 500-kg bombs. (PAF)

(Above and Next) The PAF acquired 13 reconnaissance capable Mirage IIIRPs in two batches, these also came into service of Number 5 Squadron, and except the fire control radar had the same capabilities as the Mirage IIIEPs. (Dassault)

INDIAN AIR FORCE AIRCRAFT CLAIMED BY THE MIRAGE IIIEPs
OF NO. 5 SQUADRON, PAF, DURING THE DECEMBER 1971 WAR.

Date	IAF Type	Weapon	Remarks
4 Dec.	1 Canberra	AIM-9B	Night interception over Sakesar, pilot Flt. Lt. Naem Ata
5 Dec	2 Hunters	AIM-9B	Near Lahore, pilot Sqn. Ldr. Farooq Umar
5 Dec	1 Hunter	AIM-9B	Near Lahore, pilot Flt. Lt. Shah
5 Dec	1 Hunter	AIM-9B	over Sakesar, pilot Flt. Lt. Safdar
6 Dec	1 Su-7	AIM-9B	Jammu area, Indian territory, pilot Flt. Lt. Salimuddin
10 Dec	2 Hunters	30 mm	Destroyed during the take-off run over the runway on Pathankot airfield
13 Dec	1 Su-7	AIM-9B	Over Narowal area, Indian territory, pilot Group Captain Haider
13 Dec	1 Dec	30 mm	Over Narowal area, Indian territory,pilot Flt. Lt. Shah

NOTE: A MATRA R530 missile was launched against an Su-7 but results were not seen as the Mirage had to break due to an enemy in its rear hemisphere.
The IAF claimed six Mirage IIIEPs shot down, although the PAF did not admit any loss.

For the fighter bomber role 28 Mirage 5PAs were acquired in 1970, which actually equip Nos 9 and 22 Squadrons as well as the Combat Commanders School. All the Mirages are overhauled and upgraded at the PAF's own Mirage Rebuild Facility at Kamra. (Dassault)

The most advanced Mirage delta in PAF service were the twelve Mirage 5PA3 acquired in 1979, these are equipped with an advanced nav/attack system including an Agave fire control radar capable of operating the Aérospatiale AM.39 Exocet anti-shipping missile, as illustrated in the picture, serving with No 8 Squadron at Masroor. (Dassault)

MIRAGE DELIVERIES TO PAKISTAN

"Blue Flash One"*

	s/n	Contract	Qty
Mirage IIIEP	101 to 118	1967	18
Mirage IIIDP	301 to 303	1967	3
Mirage IIIRP	201 to 203	1967	3
"Blue Flash Two"			
Mirage 5PA	401 to 428	1970	28
Mirage IIIDP	304 to 305	1970	2
"Blue Flash Three"			
Mirage IIIRP	204 to 213	1975	10
"Blue Flash Four"			
Mirage 5PA2	429 to 446	1979	18
Mirage 5PA3	447 to 458	1979	12
Mirage 5DPA2	306 to 307	1979	2
Mirage IIIO	(ex-RAAF)	1990	43
Mirage IIID	(ex-RAAF)	1990	7

Note: The PAF has acquired 40 second-hand Mirage IIIE/BEs from France, the first six were delivered in August and the rest should arrive over the next 18 months. Also, Pakistan is negotiating with Dassault for the acquisition of an unspecified number of Mirage 2000-5s.

Mirage IIIEP number 109 while on an acceptance test flight at Daussault's Bordeaux plant. (J.Cuny)

PAKISTAN AIR FORCE SQUADRONS EQUIPPED WITH MIRAGES 1995

No 5 Squadron	Rafiki	Mirage IIIEP Mirage IIIO Mirage IIIRP/RP2
No 8 Squadron	Masroor	Mirage 5PA2/PA3
No 9 Squadron	Rafiqui	Mirage 5PA
No 22 Squadron	Rafiqui	Mirage IIIDP/DPA PA/DPA2/D
C.C.S. (Combat Commanders School)	Sarghoda	Mirage 5PA

Mirage IIIDP at Bordeaux. In total the PAF received seven two-stick Mirages from France. (J.Cuny)

In 1990 the Royal Australian Air Force sold Pakistan 50 Mirages, comprising 43 single seat IIIO(A)s and seven two-stick IIIDs. These ex-Australian Mirages are being upgraded and overhauled at the Mirage Rebuild Factory in Kamra, being issued on completion of the work to Number 5 Squadron, although about half have been broken down for spares. (RAAF)

8

South African Mirages and Cheetahs

The South African Air Force (SAAF) has been one of the happiest customers of the first two generations of the Mirage, plus an indigenous development (with Israeli help), the Atlas Cheetah, has also subjected them to optimal conditions in the several operations launched against Angola, where they proved their capabilities and ruggedness. At one time they had to fly over one of the most sophisticated air defence networks ever created, with a mix of mostly Soviet SAM variants, together with rapid-fire radar-guided guns, as well as modern interceptors like the MiG-21 and MiG-23. That they came away with such few losses is a credit not only to the excellent quality of SAAF air and groundcrews, but also to the aircraft themselves. Today the first generation Mirages have been phased out from service or converted into Cheetahs, while the second-generation Mirage F1s have also suffered cuts, in 1992 the fighter F1CZ were withdrawn from service, while the ground-attack F1AZs, intended to be withdrawn in 1995, will continue,in service with 1 Squadron, according the Design 2000 plans of the SAAF until the year 2005, while 2 Squadron is in charge of Cheetah operations.

THE FRENCH MIRAGE III VARIANTS

After evaluating the Mirage III in France, the SAAF placed an initial order in 1962 and eventually in the following years received 58 new-built aircraft. These comprised 16 Mirage IIICZ interceptors received between 18 December 1962 and 9 March 1964; three IIIBZ trainers, between 14 December 1962 to 9 December 1964; four reconnaissance Mirage IIIRZs from 24 November 1966

Along the Mirage IIICZs came three IIIBZ twin-stick operational trainers which provided future Mirage pilots the means of an easier operational conversion. (SAAF)

The South African Air Force (SAAF) has been a Mirage operator since 1962. First order was for 16 Mirage IIICZ interceptors in 1962 which equipped 2 Squadron "Flying Cheetahs", these Mirages were kept in service until 1990 when they were withdrawn from use. The picture shows a five-ship formation during the early sixties in their natural metal scheme with a red cheat line along the fuselage. (Dassault)

Seventeen multirole Mirage IIIEZs were delivered to the SAAF, serving with 2 and 3 Squadrons as well as 85 Air Combat School, at least eleven of them were converted to Cheetah E standards. (Dassault)

The SAAF received three Mirage IIIDZs and eleven III2Z twin-stickers, the later had the Atar 9K-50 engine in place of the 9C-3, which had almost 2.000-lbs of additional thrust. Although classed as trainers, mostly serving with 85 ACS, they also served in operational roles during the Bush War in Angola and Namibia, the photo shows a IIID2Z in company with a IIIEZ, both carrying the JL100 combined fuel tank/rocket pod (SAAF)

to 22 February 1967; 17 multirole Mirage IIIEZs from 2 February 1965 to 15 March 1972; three IIIDZ trainers, 4 July to 18 September 1969; 11 IIID2Z trainers and four reconnaissance IIIR2Z. The latter fifteen were the first production Mirages with the Atar 9K50 engine. Remaining Mirage IIICZs and IIIBZs, of which only two were lost in accidents, were phased out of service in October 1990. A Mirage IIICZ is kept in flying status for displays and airshows.

INDIGENOUS DEVELOPMENT, THE ATLAS CHEETAH

In July 1987 Atlas disclosed a considerably modified Mirage renamed Cheetah and which was developed with considerable Israeli assistance. Conversion has affected 23 airframes of the later SAAF Mirage IIIs, —except the CZs, BZs and IIIRs,— as well as at least 45 IAI Kfirs, sold in a contract signed in 1988 for $1.7 billion, which included the airframes, as well as kits of avionics and technical support. These have been converted in South Africa to produce 15 Cheetah D trainers, 16 Cheetah E strike fighters, one Cheetah R and 36 multirole Cheetah Cs. The last is the most advanced variant, equipped with an Elta air—to-air/air-to-ground radar, Elbit mission computer and Elisra electronic warfare system, all have been produced from Kfirs, taking new serials starting at 341, this variant entered operational service with 2 Squadron in 1994. The Cheetah C is also powered by Atar 9K50 engines and like the previous variants, has an inflight refueling probe fitted to the starboard side of the fuselage. Regarding the previous Cheetahs, at least four two-seat Ds and five single seat Es were produced from Kfirs, as their serial numbers (858-861 and 862-866) are above the last Mirage III serials (see box #1). Eleven Cheetah Ds were produced from Mirage IIIDZ/IIID2Z airframes (840-847, 849, 852 and 853). The number of Cheetah Es rebuilt from Mirage IIIEZs is also 11 (822, 824, 825, 827-834), with the exception of the solitary Cheetah R (855 ex Mirage IIIR2Z).

The most noticeable features of the Cheetahs are the Kfir-style canard surfaces, as well as the similar nose on the Ds and Es, while the leading edge of the wings have the dogtooth modi-

This Mirage IIIEZ is kept in flying status by the SAAF Museum, and in 1995 was painted in this striking scheme to celebrate the SAAF's 75 Anniversary. (Lert).

(Above and Below) The SAAF operated eight reconnaissance capable first-generation Mirages, four IIIRZs and four IIIR2Zs, the later were equipped with the higher powered Atar 9K-50 engine and the doppler navigation radar, identified by the bulge below the cockpit, in the first picture an IIIR2Z is seen in the lizard scheme with a sharp (*shark?) mouth painted while serving with 2 Squadron, in the second a later grey overall machine. (Lert)

In April 1975, 3 Squadron was equipped with the first of 16 Mirage F1CZ interceptors, and in their seventeen-years operational life were heavily involved in combat operations, achieving two air-to-air kills against Angolan MiG-21 Fishbeds. 3 Squadron was disbanded on 30 September 1992 and its Mirages withdrawn from use and stored. (SAAF)

fications. (Actually a Cheetah R 855 is testing the so called "combat wing" with fixed, dropped leading edges and wingtip missile launch rails). There are additional weapons hardpoints,

plus an inflight refueling probe. The Cheetahs can carry all the indigenously developed weapons, including the V3B Kukri and V3C Darter IR AAMs, while as in the previous Mirages, internal armament is the usual two DEFA 30mm cannon.

MIRAGE III AND CHEETAH SQUADRONS

2 SQUADRON

The "Flying Cheetahs" were based at Waterkloof when the first Mirage IIICZs were received in April 1963 replacing the Sabre Mk.6s which was their previous mount. Strength was increased with the arrival of the multirole Mirage IIIEZs between July 1965 and August 1966, but these were transferred to 3 Squadron shortly afterwards. A recce capability was incorporated with the arrival of Mirage IIIEZs in November 1966, while the 9K50 powered IIIR2Z arrived in 1974. The squadron moved to Hoedspruit in 1975. When the Mirage IIICZ/BZs were phased out in 1990, Cheetah Es and Ds were received, and with the disbandment of 5 Squadron and 89 combat Flying School, 2 Squadron remained as the only Cheetah user in the SAAF, moving to AFB Louis Trichardt, and in 1994 started receiving the radar-equipped multirole Cheetah C. 2 Squadron is the largest combat squadron within the SAAF, while one of its six flights is in charge of operational conversion to this fighter.

The wasp insignia of 3 Squadron, a unit which had been established in the summer of 1940.

3 SQUADRON

Reactivated in August 1966 when it took over 2 Squadron Mirage IIIEZs, operating them until 4 April 1975 when it received the first Mirage F1CZs, passing its IIIEZs, IIIDZ and IIID2Zs to 85 Air Combat School.

5 SQUADRON

Converted to Impala Mk.IIs in 1981, 5 Squadron was relocated to the newly constructed Louis Trichardt air base on 1 January 1988 and reequipped with Cheetah Ds and Es, being declared combat ready three months later. In 1992 the squadron disbanded and its aircraft were passed to 2 Squadron.

Mirage F1CZ during a test flight out of Bordeaux, France, still without the early lizard camouflage scheme specified by the SAAF. (Dassault)

Atlas Aviation performed overhauls for most of the SAAF's diverse fleet of aircraft and helicopters, here is seen a Mirage F1CZ during such work. (Atlas)

After 3 Squadron disbanded, Mirage F1CZ 203 "Le Spectre" (The Ghost), in which Major Johan Rankin achieved his second MiG kill was sent to the SAAF Museum. (SAAF)

Mirage F1CZ of 3 Squadron loaded with two wingtip Magics and two underwing Kukri AAMs late in its operational life in the two-tone grey scheme. (SAAF)

85 AIR COMBAT SCHOOL

Formed on 1 August 1972, 85 ACS received Mirage two seaters and Impalas one year later and Mirage IIIEZs in 1975 as well as the higher powered Mirage IIID2Zs, these were kept in operations for over a decade, when the Cheetah program started and the unit, renamed 85 Combat Flying School was solely equipped with Impala Mk Is and IIs.

89 COMBAT FLYING SCHOOL

Was activated on 1 July 1986 receiving Cheetah Ds as well as some Es, the unit disbanded in 1992 and its Cheetah operational conversion was passed to a flight of 2 Sqn.

MIRAGE III SERIALS

		qty
Mirage IIICZ	800 to 815	16
Mirage IIIBZ	816 to 818	3
Mirage IIIEZ	819 to 834 & 842	17
Mirage IIIRZ	835 to 838	4
Mirage IIIDZ	839 to 841	3
Mirage IIID2Z	843 to 853	11
Mirage IIIR2Z	854 to 857	4

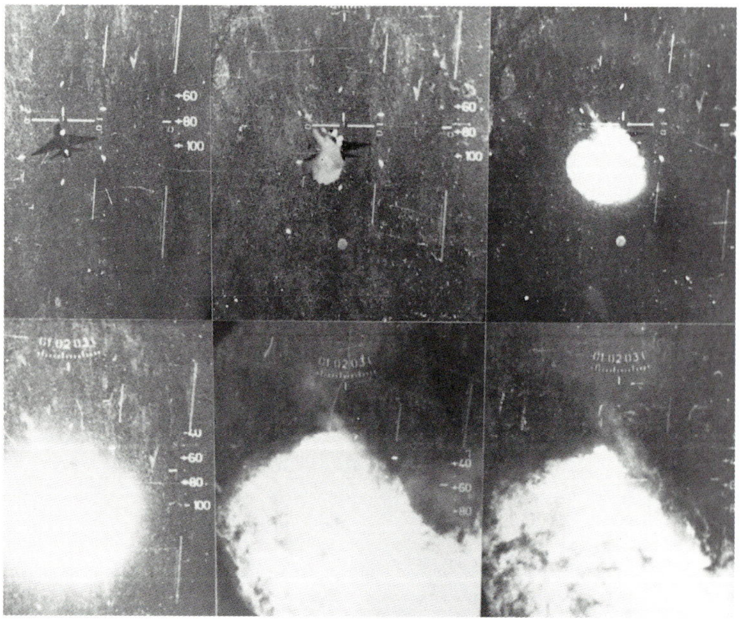

Gun camera sequence of Major Rankin's second kill on October 5, 1982. (SAAF)

CHEETAH E SPECIFICATIONS
Wing: Span 26 ft 11,6 in; aspect ratio 1.94; area 34.80 mm; canard foreplane span 12 ft 3 in; canard foreplane area 17.87 sq ft.
Fuselage and tail: length including probe 51 ft 4.25 in; height 14 ft 11.25 in; wheel track 10.6 in; wheel base 15 ft 11.7 in
Fuel and load: internal fuel 603 US gal; external fuel up to two 449, 343, 291 or 165 US gal drop tanks; maximum ordnance about 8.880 lb.
Powerplant: one SNECMA Atar 9K-50 rated at 11.023 lb dry and 15.875 with afterburning.
Speed: Maximum level speed in clean configuration at 40.000 ft 1.265 knots; maximum cruising speed at 36.000 ft 516 knots.

1 SQUADRON, MUD MOVERS

February 1940 was the official date that the squadron was established, while Major N.G. Niblock-Stuart was the first Officer Commanding. 1 Squadron earned an outstanding reputation in World War II and this is clearly evident from the honors on the squadron colors. During the War, 1 Squadron pilots were credited with 165 1/2 enemy aircraft destroyed, 26 probably destroyed and 130 damaged during aerial combat. 60 1/2 aircraft were destroyed on the ground. Aircraft operated during this period include Gauntlets, Gladiators, Hurricanes and Spitfires.

The squadron's Crusader emblem was given in recognition of the support given to the 8th Army in North Africa and the role played by the squadron to ensure air superiority, this was done by Field Marshal Montgomery who granted the unit the right to use the Crusader-Cross emblem of the 8th Army in the squadron badge. After the war the squadron was demobilized, being reformed at Air Force Base Waterkloof with Harvard and Spitfires in 1946, continuing its front line service since then. In 1950 the unit was the first SAAF squadron to be equipped with Vampire jets, which in turn were replaced by Canadair Sabre Mk.6s in 1956. Captain John Inglesby, a 1 Squadron pilot was also the first man to fly supersonic over South Africa. Late in 1963 1 Squadron became the SAAF's sole Sabre operator when it received all those formerly flown by 2 Squadron, which had reequipped with Mirage IIICZ fighters. In 1967 1 Squadron moved from Waterkloof to Pietersburg where it was used as a ground support squadron. The unit continued to operate the Sabres until 1975, while Impala Mk.1 were also flown during this period. In February 1976 the squadron was reformed at Waterkloof with the new Mirage F1AZs, a total of 32 (216 to 247) had been acquired from Dassault. This variant was quite different from the straight F1C, as the Cyrano IV radar was removed and in its place was fitted with a much smaller Aïda ranging radar, as well as a laser ranging set. The removal of the Cyrano permitted the installation of a fully retractable air-to-air refueling probe. At the same time France granted a license to build under license in South Africa by Armscor, the Mirage F1, as well as the Atar 9K50 engine, only used to build Atars for the Cheetahs. In 1978 was again declared combat ready, while three years later relocated to AFB Hoedspruit.

A Mirage F1CZ kicks-in its afterburner for quick acceleration. SAAF F1 received an interesting modification in the self-defense system, the underfuselage stability keels were replaced by others, slightly thicker, which housed flare/chaff dispensers, improving their survivability against radar-guided and IR-guided missile threats. (Lert)

Portrait of Colonel (now Brigadier) Johan Rankin, which in 1994 took charge of the command of Air Force Base Hoedspruit (SAAF)

Mirage F1AZ leaving condensation trails as it pulls in a high-g climb in afterburner. In 1995 some 22 fighter-bomber Mirage F1AZs remain in service with 1 Squadron at Hoedspruit from the 32 machines delivered. (Lert)

With its cruciform brake chute deployed a 1 Squadron Mirage F1AZ lands at its home base. According the "Design 2000" plans of the SAAF, this fighter will remain in service at least until the year 2005, forming, together with the newly delivered Cheetah C the first line fighter element of the SAAF. (Lert)

Mirage F1AZ leaving its shelter at Air Force Base Hoedspruit. (Lert)

As the main ground attack squadron of the SAAF, the unit distinguished itself during operations from South West Africa (Namibia) into Angola. The Mirages participated in operations such as Protea, Smokeshell, Sceptic, Askari, Weldmesh, Modular, Hooper and Packer. Despite the impressive defense network established by the Soviets in Angola, the squadron, employing very low level tactics —in one instance a SAM was launched against a two-ship formation which flew as low as 30 ft above ground level, and the missile exploded against a tree!— penetrated with success in Angola in all the missions it flew. Forward detachments of up to twelve Mirages were made to Grootfontein air base, near the border with Namibia, while from 16 September 1987 to 25

SAAF Mirage F1AZs received this attractive camouflage scheme, ideal for low level flying over their operations area. Mirage F1AZs have a retractable air-to-air refueling probe, while they lack the Cyrano IV fire control radar, in their place is an Aïda ranging radar associated to an advanced nav/attack system, which includes a laser spot tracker seen below the nose. (Lert)

A 60 Squadron Boeing 707 streams its three hoses as three Mirage F1AZs close-in to refuel. (Lert)

Single-seat Cheetah E strike fighters were produced by eleven Mirage and five Kfir conversions, serving with 89 CFS, 5 Squadron and 2 Squadron, but late in 1994 were withdrawn from use and stored. (SAAF)

The Cheetah's first variant was the twin-stick Cheetah D which was produced from conversion of the surviving Mirage IIIDZs, D2Zs and four Kfirs, to produce 15 machines, which serving with 89 Combat Flying School and 5 Squadron, both now disbanded, actually serve with the training flight of 2 Squadron. (SAAF)

March 1988 saw the heavies operational usage of 1 Squadron detachment from that base, which comprised 638 combat sorties flown, including attacks with smart bombs (laser guided), against the loss of two Mirages and its pilots. The first was on 20 February 1988, after a ground attack mission by four Mirages (call-sign Airwolf), two SA 13 were fired and one exploded against the number two of the first pair. The second loss took place on 19 March 1988, when the aircraft and its pilot disappeared while returning from a night attack sortie. Also several inconclusive engagements were recorded against Angolan MiG-23s. In all at least 100 SAMs were fired against the Mirages of 1 Squadron, which also had to deal with the total lack of friendly radar coverage. Various squadron members were awarded commendations for service excellence during this period, including an almost 100% readiness rating for the fleet as proof of the dedication and motivation of the squadron's logistical section.

On 30 March 1990 the squadron celebrated its 50th Anniversary and the highlight of the ceremony was the receipt of the SAAF's Operational Efficiency Award for the period November 1987 to October 1989.

Today, still based at Hoedspruit, and after the winds of its disbandment have calmed, 1 Squadron continues to be a first line fighter unit, forming with 2 Squadron (Cheetahs) and 8 Squadron (Impala Mk 1 and 2) the whole combat element of the new "core" SAAF. The unit's complement is 22 Mirage F1AZs, twelve pilots and another 138 personnel. Current Officer Commanding is Lieutenant Colonel J. Mienie, who took office in January 1995. The Mirage have received progressive upgrades since entering SAAF service in 1976, including the installation of new RWR (Radar Warning Receiver) sets, the replacement of the Magic short range AAM by the indigenous Kentran V3C missiles, the fitting of new underfuselage keels, constructed by ATLAS, which have inside flare/chaff dispensers, etc. Also one aircraft has been taken on as a development vehicle for a major upgrade program, which included a "glass cockpit" and a new nav/attack system, flying for the first time during late 1994. Also in a parallel program, an F1CZ and an F1AZ are being used explore the aircraft's capabilities reengined with one Klimov RD-33 engine (the one which powers the MiG-29). One F1CZ airframe was shipped to Russia for com-

The Cheetah D has a new nav/attack system (as well as the E) based upon the one of the Kfir C7, and actually is used as a lead in trainer for future Cheetah C pilots of 2 Squadron. (SAAF)

patibility trials. While the above two Mirages are being test flown in South Africa with the engine which is known in this program as SMR 95. This being a more powerful engine than the Atar 9K50, 200 kg lighter and has a better specific fuel consumption rate, and if the tests prove successful, it could reengine the Mirages as well as the Cheetahs.

3 SQUADRON, MIG KILLERS

The Squadron was established at Waterkloof Air Station during August/September 1940 with Major L.A. Wilmot as Officer Commanding and equipped with Hawker Hurricane fighters. The unit was divided into a sea party and an air party, the latter travelling in Vickers Valentias and acting as a servicing flight to the Hurricanes on the way up to East Africa. The sea party left from Durban on 15 October 1940 aboard the "Durban Castle", and after an uneventful voyage disembarked at Mombasa eight days later, when they immediately entrained for Nairobi. The squadron flew its first sortie on 28 October 1940 but it was not until 22 November that Lt Allen drew first blood by destroying an Italian Caproni CA 133.

3 Squadron's famous crest, "The Fighting Wasp" with the motto Semper Pugnans (They Always Fight), was adopted on 1 December 1940. The unit was in Italy flying Spitfire IXs when the war in Europe ended, having attained an impressive record in almost five years of combat. After the conflict the squadron was disbanded, but on Saturday 6 September 1952, this fighter outfit again came into existence with Major Dan Till as Commanding Officer, until October 1959 when the it was again disbanded. Then in August 1966 it was once again reformed, equipped with Mirage IIIEZs, as a flight under command of the O.C. 2 Squadron until April 1975. On 20 February 1970 colors were presented to the squadron in a ceremony at Air Force Base Waterkloof.

In April 1975, 3 Squadron was equipped with the first of 16 Mirage F1CZs (200 to 215) under the command of Commandant P.B. Huyser. During its service life the Mirage F1CZs participated in operations against the SWAPO such as Sceptic, Protea, Daisy and Meebos. A total of about 500 operational sorties were flown without loss of any aircraft. Also 3 Squadron has participated in 271 combat sorties of various operations, different to those related above, in the Operational Area, which included two MiG-21 kills (narrated below.) Due to the important scale down of the SAAF in the period 1990-1994, 3 Squadron was disbanded on 30 September 1992, while the surviving Mirages were placed in storage. Concerning losses, the unit lost its first two F1CZ in a collision in February 1979 (203 and 204), but their airframes were stripped for parts to equip two other airframes with the same serial numbers which were produced by Dassault in France. The same was done in South Africa with 205 which had crash landed on 8 February 1985 and 206 which had been badly damaged by a SAM over Angola on 27 September 1987, making a new 205. Other losses, not replaced, were 200 on 15 February 1979, 208 on 4 November 1980, 214 and 215 on 28 December 1987. Thus at its disbandment ceremony, 3 Squadron presented its eleven surviving Mirage F1CZs.

Dramatic painting of two Cheetah Ds during a practice strike mission loaded with inert 500-lb bombs, while in 89 CFS service. All Cheetahs carry fixed air-to-air refueling probes on their right side. (Tiro Vorster)

The Cheetah C is SAAF's newest combat aircraft having been surrounded in secrecy; thirty-six have been delivered, and most of these serve with 2 Squadron, this differs from earlier variants in having an Elta EL/M-2032 multimode pulse doppler radar which gives an entirely nose shape, while also has a new frameless windshield and other improvements, all were produced by conversion of Israeli Kfirs. The picture shows Cheetah C number 342 wearing a special color scheme to celebrate the SAAF's 75th anniversary. (Barnard)

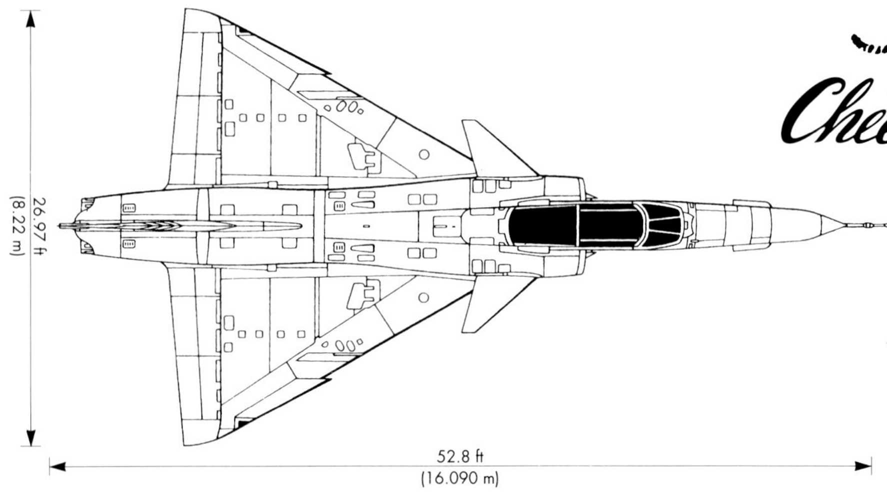

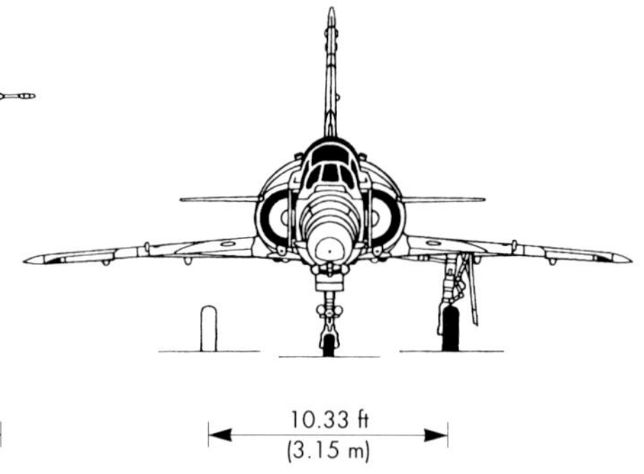

MISSIONS
- Intercept
- Combat Air Patrol
- Ground Attack
- Training

ARMAMENT
- Two 30mm Cannons
- Bombs
  - Mk 82 500lb
  - Mk 83 1000lb
  - Cluster
- A/A Missiles

CREW
- Pilot
- Navigator

FUEL CAPACITY
- Normal fuel capacity 3290 Lts
- Maximum fuel capacity 6570 Lts

AVIONICS
- Weapon delivery and navigation system (WDNS)
- Radar
- Glass Multi function cockpit displays
- Electronic warfare system
- Communication system

MASS
- Empty weight 7 340 kg
- Maximum weight 13 600 kg
- Wing span 8,22 m
- Height 4,5 m
- Length 16,18 m

FLIGHT PERFORMANCE
- Maximum speed Mach 1,8
- Cruise speed 540 kts
- Landing speed 160 kts

POWER PLANT
- Atar 09 series

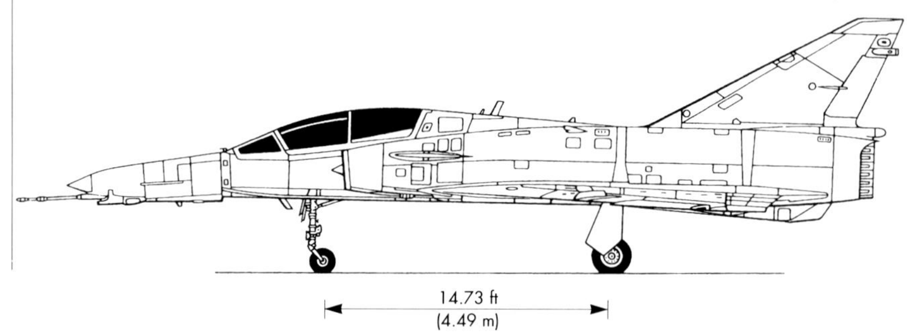

Three view drawing of the Cheetah D. (Atlas)

MAJOR RANKIN'S DOUBLE

The Mirage F1, a fast and deadly fighter and perhaps one of the most effective warplanes in the world, has been used in combat by all of its eleven customer air forces, except two: Jordan and Spain, but arguably where this sleek fighter was better used is in the various wars it participated in while serving with the South African Air Force, even in 1981 and 1982, two air-to-air kills were recorded by SAAF F1CZs from 3 Squadron over Angolan MiG-21s flown by Cuban pilots.

Operation Daisy was launched on 1 November 1981, although it wasn't the first time the SADF (South African Defence Forces) had directly intervened in Angola to prevent the SWAPO forces from crossing the border into South West Africa (Namibia). Less important than previous operations, Daisy lasted only 20 days. Its primary target was to strike Angola and make the SWAPO men understand that this was not a safe sanctuary. As the operation developed and the South African Army occupied SWAPO camps and destroyed large quantities of equipment and ammunition, the MiG-21s of the well equipped Angolan Air Force tried to intervene in order to dispute the SAAF's air superiority. Brigadier Johan Rankin, presently officer commanding Air Force Base Hoedspruit, then a Major serving with 3 Squadron flying Mirage F1CZ, continues with the narrative:

"...It was 07:00 local on November 6 (1981), and I was on cockpit stand-by with my wingman, Captain Johan du Plessis at the advanced Ondangwa Air Base. At this time in the morning the temperature was still bearable, permitting us to stay alert in the aircraft and even to reduce the time it would take for an interception. It was then that our radars detected hostile aircraft heading towards the area occupied by our forces close to Cahama. In the past, the MiGs had always slipped away when we tried to engage them in combat, so we decided to teach them a lesson. We used a land line to discuss with the radar controller a low level penetration up to Cunene River to avoid detection by the Angolan's radars, where after we would climb back into radar cover and establish radio contact for the ground controlled interception. (GCI) We scrambled immediately and followed the profile agreed upon. Our radar controller kept us informed about the position of the MiGs until we lost radio contact. On crossing the Cunene, we accelerated to combat speed and started a climb to 25,000 ft. The climb, which did not take more than thirty seconds, was not detected by the MiGs and had without a doubt caught the Angolan radar operators unawares. Our radar controller detected us immediately and as we re-established radio contact, he reported the MiGs approaching us head on but offset approximately five nautical miles to our left. My wingman was the first to see the two

MiGs which appeared in front of us, in our ten o'clock at about five nautical miles. They were flying in a typical Warsaw Pact formation, the wingman trailing his leader by about 1,500 meters. We jettisoned our external tanks and entered a hard left turn, which placed us behind the MiGs. We then found ourselves facing the sun, which persuaded us not to use our MATRA 550 Magic infrared guided missiles —they could have locked on the sun instead of the MiGs. I approached the first MiG which was in formation with his wingman. I fired a burst of explosive shells from approximately 350 meters. His centerline external fuel tank MiG started losing fuel, indicating he had been hit. The MiG then made a tight turn to the left and I followed, holding off at approximately 1,000/1,500 meters behind and to the right of it. During this time my wingman followed the leader who executed a split-S to the right hand side and down (a typical Warsaw Pact escape maneuver). My MiG started a right hand turn, which was going to place him behind my wingman. I therefore entered a curve of pursuit on it until I reached 500 meters when I opened fire with cannons again. The MiG exploded virtually immediately, breaking into two pieces behind the cockpit. I looked for my wingman, as I was worried that he was chasing the lead MiG deeper into Angola, beyond a position where I could cover him. When I looked back at my MiG, I noticed the pilot's parachute about 10,000 ft below me and the rear section of the MiG spinning down.

I started diving down towards the position where I had last seen my wingman and his MiG, but could not find them. At this stage, our radar warned us that there were other MiGs on their way to the sector. I called back my wingman and we returned to Ondangwa...."

Major Rankin's aircraft was number 213. Several days afterwards the Angolans found the body of their pilot. For the SAAF fighter pilots it was a time of celebration, but also of worry, Angola was receiving more advanced MiGs both of the Fishbed and Flogger variants, and in a near future could contest the air superiority the SAAF had obtained. About a year later, on 5 October 1982, through one of those coincidences which surround the aviation world, Major Rankin found himself in the center stage of a dogfight between two Mirage F1CZ and two MiG-21s. The SAAF was undertaking reconnaissance missions over the SWAPO bases, aware that this organization was preparing an attack towards the border territories of South West Africa, he follows:

".....Together with my wingman, on this occasion Captain Cobus Toerien, I escorted a Canberra on a photo recce mission. We were once again near Cahama, this time southwest of it and flying at 20,000 ft when we were warned by our mission controller on the ground that MiG aircraft were on their way to intercept us. As the Canberra is relatively slow and vulnerable, I immediately sent it back to Ondangwa and requested the controller for intercept instructions. We had to chase the MiGs away, or engage them to give the Canberra sufficient time to get away successfully. We were instructed to climb to 30,000 ft and turn onto a northern heading, we accelerated to Mach 0.95, which was the limiting speed for our external fuel tanks. We were immediately told that the MiGs were 12 nautical miles away, approaching at supersonic speed and would cross on our right hand side. I saw the MiGs at approximately five nautical miles at the two o'clock position and at the same level as us. They were on exactly the flight path our controller had predicted. We jettisoned our external fuel tanks, went into afterburner and started a hard right hand turn in pursuit. I don't know at which moment they located us, but when we passed, they fired their missiles. I also don't know whether they acted on orders from the ground or whether they thought that this would render their airplanes more manageable. Maybe they thought that they could hit us with IR-guided missiles, probably Atolls. The firing of these missiles will always remain a mystery to me. We knew, by intercepting radio communications, that their radar on the ground had detected us. Maybe the MiG pilots, guided by their GCI controllers, thought they could shoot us down without engaging in a dogfight. The missiles had no effect. We completed our turn through 180-degrees, while the MiGs only did a gentle turn to the right. I got the impression that they were maintaining supersonic speed, which decreases the MiG-21 turn rate tremendously. This was confirmed by the fact that they were opening the distance on us. They were slowly turning through southwest to west, but could still threaten the Canberra. We tried to catch up with the MiGs, but realized that we would not be able to close the distance. I switched my aircraft's intercept radar onto transmit, hoping that the MiG pilots' radar

Right: Oil painting by South African artist Tiro Voerster of Major Rankin's second MiG-21 kill on 5 October 1982, on that day he was flying Mirage F1CZ number 203, nicknamed "Le Spectre" due to its new grey scheme.

Oil painting by artist Tiro Vorster showing the SMR-95 re-engined Super Mirage F1AZ flown by SAAF test pilot Major Johann "Blokkies" Joubert, in a vertical climb during the international air show at Waterkloof air base, as part of the SAAF 75th anniversary celebrations in October 1995. Note the pilot tube is painted orange to indicate that it contains test instrumentation, and the redesign of the exhaust nozzle for the Russian rebuilt engine.

warning receiver would warn them and force them to turn into us, allowing us to close the distance. Whether it was for this reason, or because they had received orders from the ground, I don't know, but they reversed their turn to the left, returning to the heading on which the Canberra was running away. I immediately switched the radar off again, cut the corner and went into an energy acceleration until we reached about Mach 1.3 as at this speed the Mirage F1 outruns the MiG-21, which, on the other hand, is better at subsonic speed. I found myself in a position to enter a curve of pursuit and although the radar had not locked on again I fired a Magic infrared missile when the range appeared correct. It tracked the trailing MiG until its motor burned out and then lost its lock-on. Subsequent examination of the gun camera film showed that the missile had been fired at 3,000 meters, that was the extreme limits of its range for the situation. The missile was lost.

Again I went into an energy acceleration to close on the MiGs and, at about 1,500 meters from the rear one, I fired my second Magic which tracked the MiG which entered a split-S. The missile exploded right behind it. The MiG was hit but still controllable and continued its left hand turn to return to base, trailing smoke. According to information received afterwards, the MiG was severely damaged and was not able to lower its landing gear, requiring a forced landing which caused serious damage to it.

After the first MiG had fled and because my wingman was still trailing me by some distance, I closed in on the second one (the leader). The MiG started to roll over to the left to enter a split S with me following. I was closing rapidly and at about 230 meters I started firing the DEFA cannons. The MiG exploded right in front of me and I could not avoid the fireball of the explosion. My aircraft's engine developed a compressor stall due to the explosion and I pulled up to clear it.

Initially it would not clear, but after cutting the engine and doing a hot relight, it cleared. Captain Toerien, who had been following me throughout this episode, formed up on me and inspected my aircraft for damage while we turned south to head for the border and Ondangwa. Captain Toerien had not been able to fire any missiles because of his fear that they could have been locked onto my aircraft. He had been satisfied to cover me instead.

As in the case of the first fight in 1981, I think that the MiG-21 pilots had been Cuban. Intelligence supported this assumption to be almost a fact. Whether they were Angolan, Cuban or Russian the MiG pilots in Angola were all flying according to the rigid operational doctrine used by the Warsaw Pact. No room was left for individual initiative. Ground control gave the orders and the pilot obeyed. It was evident that, even although some pilots were more talented than others, they were all more or less identical in their behavior..."

On that day Johan Ranking was flying Mirage F1CZ number 203 nicknamed "Le Spectre" (French for Ghost) by its pilots because of its two-tone grey camouflage, and its effectiveness, which then was under evaluation. This aircraft is displayed today at the SAAF Museum at Swartkop. The air combats of 1981 and 1982 were without doubt the only ones where Mirage F1s and MiG-21s had been engaged in battle.

Brigadier Rankin received his pilot dispatch in 1968, his first posting was with 42 Squadron flying Cessna 1852 lightplanes, recording 800 flight hours in the operational area. Later he qualified as an instructor on Harvards and later Impalas (license built version of the Macchi MB326). In 1977 he joined the 85 Combat Flying School which was then equipped with Mirage IIIs and Impalas. In 1980 he had already recorded 20 operational missions on the Mirage III, and in the same year he was posted to 3 Squadron flying Mirage F1CZs. He completed 50 combat missions with the French fighter claiming the two kills already related. From 1984 to 1986 he was the C.O. of the Jet Flying School based at Langebaanweg, and from there joined 1 Squadron, equipped with Mirage F1AZs, in which he flew another 80 combat missions in Angola. Brigadier Rankin was assigned since 1988 to several Embassies in Europe and as a staff officer at SAAF headquarters in Pretoria, while in 1994 he took charge of the command of Air Force Base Hoedspruit, where 1 Squadron is located, flying Mirage F1AZs and 85 Combat Flying School equipped with Impala Mk.I and IIs as well as the Silver Falcons aerobatic team flying Mk.Is.

A pair of Cheetah Cs of No 2 Squadron, the especially painted "Bogey" Cheetah in the foreground has a special scheme for the SAAF's 75th Anniversary celebrations in 1995. Note the air refueling instalation of this fighter, which can be claimed as a second-generation Mirage III. (SAAF)

The insignia of No 2 Squadron, the "Flying Cheetahs", SAAF, actually equipped with Cheetah C multirole fighters.

9

Iraqi Mirages, Two Wars and Still Flying

The Al Quwwat al Jawwiya al Iraqiya (Iraqi Air Force - IrAF) has been the largest export customer for the second-generation Mirage F1. The IrAF operated them extensively during the eight-year war with Iran. Its pilots' obtained first rate combat experience, which was of little value during the "Second Gulf War" in January-February 1991, during which they were confronted by the best in the coalition. For most of them, the best course of action was to remain alive and conserve the aircraft for a possible third round.

After placing large arms contracts with the Soviet Union, Iraq turned to the West, more precisely to France for further supplies. Negotiations for the sale of 50 Mirage F1s had first begun in January 1970, when the type had just started its development phase, but were broken shortly afterwards. Renewed interest eventually led to agreement with France in June 1977 for 36 Mirage F1s, plus 36 options, and associated Super 530F1s and Magic I air to air missiles.

The first F1EQ single-seater flew on 28 May 1979, followed by the F1BQ on 6 February 1980. Deliveries began in April 1980 to CEAM (Centre des Experiences Aériennes) at Mont-de-Marsan for training. By the time the aircraft were ready to leave France, Iraq had attacked Iran and was embroiled in what was to become a long and bitter eight-year war. France declined to join the international arms embargo and allowed the first four aircraft to pass through Larnaca, Cyprus, for delivery on 31 January 1981, disguised as deliveries to Jordan. Further orders were placed in late 1979 (24), February 1982 (30), September 1985 (23) and 1988 (15). In total, Iraq ordered 110 single-seat and 18 two-seat Mirages, not all of which were delivered (see box #1 for serials, variants and deliveries).

All Mirages, like those of Libya, were unique in having an HF fillet antenna at the forward joint of fin and fuselage plus the standard BF radar warwing antennas. The series 200 aircraft were fitted with refuelling probes, while from the F1EQ-4-200 all have

The Iraqi Air Force was the largest export customer of the second generation Mirage, but what it is not very well known is that since the mid-sixties it had tried to obtain also the first generation Mirage, in that case the Mirage IIIE. Diverse factors conspired against this buy, mainly the financial state of the country, but this was solved towards the mid-seventies, and then the Mirage F1 was already available. (Dassault)

Negotiations for the sale of an initial lot of 50 Mirage F1s had first begun in January 1970, when the type had just started its development phase, but these were broken down, then in June 1977 a contract was signed for 36 aircraft, plus another 36 as options; Iraqi pilots evaluated the MIrage F1 from the development batch. (Dassault)

VOR/ILS antennas, which were retrofitted to previous variants. This variant was the first multirole Iraqi Mirage, having provision on the centerline stores pylon for the COR-2 and Harold reconnaissance pods as well as Intertechnique buddy refuelling pods. The F1EQ-5, which first flew on 7 July 1983, had Thomson-CSF Agave radar in place of the normal Cyrano IV for their anti-shipping role with Aérospatiale AM39 Exocet missiles. Some of the EQ-5s had their typical Iraqi lizard scheme of sand and brown on the upper surfaces over a blue-grey scheme. The next model was the F1EQ-6 which had several improvements over previous variants, including the BF radar warning receiver replaced by the Thomson-CSF Sherloc. This series was terminated in 1987, being followed by the F1EQ-7, of which six of the twelve on order had been delivered. In 1989 Dassault terminated deliveries when funds from Iraq were not forthcoming, also pending of delivery were three new production F1BQ two seaters, plus another three of the same variant in overhaul. These were placed in storage

The first batch of Mirage F1EQs were accepted by the IrAF in August/October 1981, arriving to Iraq when the war with Iran had raged for a full year, they were put into service and soon proved their worth. (Dassault)

A total of fifteen Mirage F1BQs two-stickers were delivered to the IrAF, which except the two DEFA 553 cannon had the whole weapons system of the single-seats, and although mostly used as combat/transition trainers, they were also used in war missions, both in the war against Iran and in the Gulf War. In fact Iraq stole the surviving Kuwaiti AF Mirage F1BK two-stickers during the invasion of the emirate, and these have not been returned. (Dassault)

following the UN arms embargo after the August 1990 invasion of Kuwait. These twelve aircraft, placed in storage, were offered to Spain in 1994, but were rejected in favor of the Mirage F1EDA/DDAs from Qatar (see chapter COMBAT VETERANS, FROM QATAR TO SPAIN).

For test flying and delivery, IrAF Mirages carried five-letter call-signs on the fin, with the last two letters repeated on the forward fuselage. Basically the first 32 Mirage F1EQs (4004-4019-4020-4035) were assigned as interceptor and for air superiority roles during the war with Iran. They claimed at least 36 air to air kills, while firing 100 radar-guided Super 530 missiles, usually in pairs, up to mid-1983, although the shorter ranged Magics and cannon were also used. Iranian types claimed were mostly F-5E/F Tiger IIs and F-4 Phantoms, but at least one Grumman F-14A was surely shot down on 22 November 1982.

Mirages also provided escort for strike bomber operations and flew reconnaissance missions with the COR-2 pod. Up to the end of 1984, some seven Mirages had been lost to Iranian defenses. Their main operating base was Qayyarah West, but two-four aircraft detachments were made to other air bases closer to the front line.

The Exocet equipped Mirages were acquired to continue the "Tanker War" in the Gulf, begun with Aérospatial SA 321 Super Frelon helicopters, followed by five loaned Super Etendards in 1983-1985. Missile software problems resulted in failure of the first Exocet-armed Mirage mission on 3 December 1984, delaying the first successful anti-ship attack until late February 1985. Slightly over 100 ships were attacked by Iraq between mid-1984 and the end of hostilities, many of them by Mirages. Exocets were credited with all but four of the 41 ships written off. The most famous raid was the accidental attack on the frigate U.S.S. Stark on 17 May 1987 which resulted in the ship being severely damaged and 37 crewmen losing their lives. Mirages also participated sporadically in the CAS (Close-Air-Support) and BAI (Battlefield-Air Interdiction) against the Iranian Army. By 1990, 20 F1EQ and

Mirage EQ fitted with a COR-2 reconnaissance pod. During the Gulf War nine Mirages were destroyed in air combat, seven by USAF and two by the Royal Saudi AF. (Dassault)

F1EQ-2, together with eleven of the fifteen two seaters, were assigned to three squadrons at Qayyarah West, Kut al Hayy East and Al Asad while the attack-configured F1EQ-4/5/6/7 were operated by three squadrons from Al Sahra, Tikrit and Tallil.

Five EQ-4s and six EQ-5s had been lost during the war against Iran, but no EQ-6s or EQ-7s, the latter delivered after the conflict. Iraqi Mirage armament was very advanced. For the air-to-air mission they had the Matra Super 530F-1s radar guided AAMs and the Matra 550 Magic I/II short range IR-guided AAMs. For the air-to-surface, Iraqi Mirages had the Matra ARMAT anti-radar missile, Aérospatiale AM39 Exocet, Aérospatiale AS30L laser-guided missile — including the Atlis designator pod — as well as Thomson-Brandt 68mm and 100mm rockets. The Russian X29L missile was also interfaced with the Atlis pod. Meanwhile, at least two mission simulators had been acquired from LMT in France, while most of the maintenance on the airframes, avionics and engines were being performed in Iraq.

HUSSEIN'S PILOT

The long attrition war, which during eight years confronted Iraq and Iran, produced a generation of pilots within the Iraqi Air Force whose professional career developed mainly while their country struggled in a conflict which never seemed to reach an end. Many pilots loyal to the regime fought, survived, or died; but there was a percentage who were only marginally committed, and they began to challenge the waste of men and resources, and the way the President was running things. In a political system like that one, this kind of opposition has only two possible ends: jail or the death squad. Many who doubted the President ended their days in this way. A few were fortunate enough to leave the country, settling as exiles under new identities in order to avoid Iraqi agents disguised as embassy officials in charge of locating and — if possible — eliminating dissidents.

This narrative, which must remain anonymous for obvious and understandable security reasons, offers the first hand testimony of one of those dissidents who went into exile. It was prepared on the first days of January 1991, when the Coalition war machinery was gearing up for the forthcoming Desert Storm operations. He knew it and he also knew that his former air force (which according to him was a powerful, but constantly pursued air arm) would fail to prevent allied air superiority. This adventure was a dead end street for the air force, a position it faced due to policies of a lunatic leader. Here are his words:

". . . Late in 1985 I was posted to one of the Mirage F1 equipped squadrons. Several months before I had been sent to Orange, France, for the conversion stage, via simulator and dual seaters. Afterwards this instruction was made entirely in Iraq when all the training aids were received, including some Armée de l'Air second flight instructors. My earlier postings had been in a Hunter squadron based on northern Iraq, where I undertook many ground attack missions against the Kurdish guerrillas and the Iranians. At this airfield I lost a fellow pilot and friend from my same flying course during a surprise attack by a section of Iranian Phantoms. I was able to take-off but, due to the Phantoms' higher speed, they escaped easily. My next posting was in a MiG-21 squadron. Soon I mastered this beautiful fighter and in the two years I was

Mirage F1EQ number 4009 while on delivery to Iraq. During the Gulf War 24 Mirages fled to Iran, and these have not been returned yet, while a similar number were destroyed at their air bases in Iraq, both inside the hardened aircraft shelters, or caught on the open. (author collection)

there, three types of missions were performed: air defence, air cover, and ground attack — although for the latter two the MiG-21 were severely handicapped by range and warload. During that period my squadron lost seven aircraft and five pilots, only one downed by the Iranian air arm, the others due to AA fire or accidents. From that era I remember some bold raids performed by Iranian aircraft. Thanks to the technical support provided by Israel — at least that's what we were told — a considerable share of the Iranian Phantom fleet was operational. These raids were made against civilian or military targets. Rarely could we catch the Phantoms, but SAMs and AA fire took a toll. During all the combat sorties I made in the MiG-21, I only received damage in one and that amounted to two small arms caliber holes on the vertical fin during a ground-attack sortie. About that time, I had several engagements with enemy aircraft but without positive results; although, in one instance I was able to shoot down an Iranian AH-1J gunship with 23mm gunfire, although this action nearly took my life, as the MiG almost stalled as I had to slow down considerably to center the helicopter on the pipper. Fortunately, I was able to recover from that situation.

I spent the next two years as an instructor in the Flying School. There a fellow officer confirmed for me what I had been hearing since a few years before: all the pilots above the rank of Captain from the Air Defence Regiment in charge of covering the Osirak nuclear reactor bombed by the Israelis in June 1981 had been shot. This madness started to eat at me. Afterwards I was selected for converting to the Mirage F1 in France. My first impression upon seeing this warplane was one of confidence; besides its good flying qualities, its nav/attack system was quite advanced.

I performed close to 300 combat sorties in the F1, and perhaps the most exciting mission was the one I made one day in early 1987, consisting on an attack against the Iranian petrochemical complex at Kharg Island. The strike was composed of four Sukhoi Su-22 with retarded bombs, two Mirages with Exocet missiles, and four Mirages used as escorts. Deploying to an advanced base on southern Iraq, the attack would take place at sunrise. I was element leader in the second covering section. Our aircraft were configured with two Magic IR-homing AAMs, full 30mm ammo and two RP35 1200-litre drop tanks under each wing.

After take-off in darkness, the ten aircraft joined up with the Exocet armed Mirages in front. The Sukhois were behind while we were on both sides of the strike gaggle. We were at 5000 ft. and they were at 2000. With the sun slightly over the horizon, we entered the Gulf and started to descend in unison. At some 30km from the island the two Exocet-carrying Mirages climbed and switched-on their Agave radars, then fired the Exocets, which initially fell ballistically before their rocket motors accelerated them just above the waves, leaving a whitish smoke trail. I lost them from sight in the morning's mist. Both Mirages made a hard 180-degree turn, going for home. The Sukhois would bomb in shallow dives and we turned left to orbit between Kharg and the Iranian coast. Now I had the island on sight and a dense smoke column rose from the harbor area, possibly a tanker had been hit by the Exocets.

The Su-22s were flying really fast toward the target with their wings fully swept back, then Iranian flak started to shoot. I could see the tracers and the little black puffs. The Su-22s dropped their lethal bomb loads and a few seconds later several explosions rocked the complex. The Sukhois started a wide right turn, heading for home. Then a large fireball blossomed just in front of the of the flight leader, who went through it; after emerging, the fighter bomber commenced rolling to the right, shedding pieces until upside down, then a wing fell off and in this attitude the stricken fighter bomber impacted against the water in a big cloud of spray. No ejection was observed. It had been shot down by one of the Hawk SAMs.

Attentive as we were to that drama, we didn't take notice of something more dangerous — for our integrity — that was closing in from our three o'clock position, almost within the sun's circle where they were nearly invisible, the perfect jump. My wingman gave the warning call — "Bandits at three o'clock!" I turned my head and saw something reflecting just to the right of the sun. We broke right to meet them head-on; now I could discern their shapes. They were a pair of Phantoms identified by their trails and size with two F-5 Tigers in trail.

Then I saw that the Phantom on my right, we were fully head-on and at about 3000 meters, launched what appeared to be a small missile, perhaps more to confuse us than from having obtained a positive lock-on. The trick worked. The first Mirage section fell behind and a little higher than our section during the right turn, and then broke to the left, easing the enemies' pursuit. The enenmy now started a left turn.

Mirage F1EQ-4 number 4526, this subvariant comprised 28 examples and was the first to include air-to-air refuelling probes. (author collection)

Photographed while on delivery, a Mirage F1EQ-5 4567. This batch comprised twenty machines, serial numbers 4560 to 4569, note the 2200-liter long range tank. (author collection)

The Phantom which had launched the missile crossed below and to my left, going like hell, although I was already performing a half roll in a high-g descending turn to gain a favorable position. Now I had him in front at some 3000 meters. He and I were both in full afterburner. I had lost from sight the other aircraft, both friendly and enemy, except for my wingman, who was in trail some 500 meters behind. The Cyrano radar showed signs of trying to lock on the F-4, very soon the HUD provided cues that I was within the shooting envelope, while the Magic AAMs started their bell-rattle that their seeking heads were locked. The Iranian pilot quickly noted his dangerous situation and without any support from his wingman, broke hard to the right. I pressed the trigger, after a brief delay the left wingtip Magic fired and, to be sure of the kill, I fired the second also.

The F-4 was now with a 90-degree bank angle and 20-degrees offset to my right and some 1000 ft. over the water. The first missile hesitated then passed by his right without detonating, the second closely followed the first, but this time the proximity fuse worked. I observed a little firewall under the Phantom's belly, followed by a trail of vaporized fuel, then a long sheet of flames. Almost at the same time both crew ejected and the F-4 went out of control. I pulled up to avoid debris, losing it from sight, but my wingman saw it crashing into the sea and both crewmen hanging from their chutes. As fuel was getting low, we returned home where we found out that everybody had landed safely except the Su-22 flight leader. I had made a kill and this automatically implied a medal, but my worst problems would arrive soon . . ."

Two months later, the trouble turned up in the form of an erroneous attack on the frigate U.S.S. Stark, with the subsequent purge of the pilot. Our anonymous interviewee protested, by the way, that the matter was handled and he was following instructions from the air base Baath Party political officer. His C.O. grounded and arrested him. This was his situation until August 1988 when the war ended. By then he could move to Baghdad with relative freedom, contacting there other young officers and becoming more and more critical with the regime. By the end of that year, and through the word of a friend, he learned that the feared internal security service — the party's secret police — was after him. Almost miraculously, some people contacted and helped him get out of Iraq. The fact that the pilot was a bachelor eased the task. After an identity change and spending some time in a Middle East country, he settled in Europe, where he lives peacefully.

According to Israeli intelligence sources, early in 1990, Iraq and Jordan proposed establishing a joint Mirage F1 squadron at H-3 air base, just over the Iraqi side of the international boundary line. In the end this proposal did not take place, although Jordanian pilots flew on exchange postings with IrAF F1 squadrons, although it is not known if they participated in combat operations.

At the start of Operation Desert Storm, the IrAF had 81 Mirage F1EQs and eleven F1BQs, of which 62 of the Mirages were operational. Since Desert Shield started in August 1990, they had not been operated extensively. Some exercises were undertaken, which served the allies by allowing them to gather lots of intelligence on the IrAF capabilities and "modus operandi." Very few raids of the Coalitions' air forces were contested by the Iraqis, preferring to conserve resources, and from 26 January 1991, to take refuge in Iran, where they were impounded. A total of 134 IrAF aircraft followed this system, of which 24 were Mirages.

Meanwhile, especially during the first part of the war, Mirages rose to challenge the allied air armada, and nine of them were shot down, six by USAF F-15Cs, one by an EF-111A Raven (through maneuvering) and two by a R.S.A.F. F-15C, the latter two attempting an Exocet strike on coalition warships in the Gulf (see Box #2 for details of air-to-air kills). Also a British R.A.F. Tornado destroyed an F1 on Al Asad air base during an attack on 17 January, while other Mirages were destroyed in LGB attacks on hardened aircraft shelters at Iraqi air bases, while the Mirages did not get any kills. According to intelligence sources, when the war ended, the badly mauled, but not destroyed, IrAF had some 28 Mirages remaining intact in their airfields, of which roughly half were operational. By 1995 the number of operational capable aircraft had grown due to the delivery of critically needed spares from a third party.

IrAF MIRAGE F1 VARIANTS

Variant	Serials	Quantity	Acceptance dates
F1EQ	4004-4019	16	26 Aug. 80-14 Oct. 81
F1EQ-2	4020-4035	16	9 Apr. 81-10 Mar. 82
F1EQ-4 (1)	4500-4503,4506-4529	28	31 Dec. 82-11 Jul. 84
F1EQ-5	4569-4579	20	19 Dec. 83-28 Feb. 85
F1EQ-6	4600-4608,4615-4623	18	14 Jan. 88- Dec. 88
F1EQ-7	4650-4651,4655-4664	12	(six delivered 1989)
			(other six not delivered)
F1BQ	4000-4001	2	27 Aug. 80-11 Sep. 81
F1BQ	4002-4003	2	17 Jun. 81-24 Jul. 81
F1BQ	4504,4505,4556-4558	5	15 Apr. 82-14 Dec. 84
F1BQ	4609-4614	6	13 Jan. 88- Jul. 89
F1BQ	4652-4654	3	(not delivered)

(1) From this variant on, all the single seats had air-to-air refuelling probes.

IrAF MIRAGE F1s SHOT DOWN DURING DESERT STORM-1991

Date	By	Wing/Squadron	Weapon	Crew
17 Jan	F-15C	33rdTFW/58thTFS	AIM-7M	Capt. R.E. Graeter
17 Jan	F-15C	33rdTFW/58thTFS	AIM-7M	Capt. R.E. Graeter
17 Jan	EF-111A	66thECW/42thECS	ACM	Capt. J. Denton (p)
				Capt. B. Brandon (wso)
17 Jan	F-15C	1stTFW/71stTFS	AIM-7M	Capt. S.W. Tate
19 Jan	F-15C	32rdTFG/32ndTFS	AIM-7M	1Lt. D.G. Sveden
19 Jan	F-15C	36thTFW/525thTFS	AIM-7M	Capt. D.S. Prather
24 Jan	F-15C	3rdFW/13thFS	AIM-9L	
24 Jan	F-15C	3rdFW/13thFS	AIM-9L	

Note: Both kills performed over the Gulf by Royal Saudi Air Force pilot Captain Ayhed Salah Al-Shamrani (call-sign "Stinger 20") on two Mirage F1EQ-5 loaded with Exocet missiles on their way to attack allied shipping. The two escorting MiG-23s escaped unscathed.

Date	By	Wing/Squadron	Weapon	Crew
27 Jan	F-15C	36thTFS/53rdTFS	AIM-7M	Capt. B.D. Powoll

(On the same mission Captain Powell also shot down a MiG-23 with an AIM-7 Sparrow.)

10

Combat Veterans, From Qatar to Spain

The arrival of the first seven Mirage F1EDA/DDAs to Manises-Valencia air base for service with 111 Escuadrón (call-sign "Dólar"), the flying component of Ala 11, was the culmination of an interesting story which began with the frustrated Mirage IIIEE/DE upgrade — canceled in July 1991. Once the Spanish Air Staff made the decision to withdraw the Mirage IIIs from service, the only thing which could be done (as nobody wished to disband Ala 11) was to reequip 111 Escuadrón with a dozen of Mirage F1CE/EEs. This solution was also frustrated, as the maximum strength that the squadron had was eight Mirage F1CEs loaned from Ala 14. In fact, this wing was the owner of the 50 surviving Mirage F1s acquired by Spain (30 F1CEs, 17 F1EEs and three F1BEs), which had to be shared by four squadrons, 111 of Ala 11, 141 and 142 of Ala 14 and 462 of Ala 46. This state of affairs could not last very long and early in 1993 the Ejército del Aire started to look elsewhere for second-hand Mirage F1s. After almost a year of evaluating the possible options it was considered that the best aircraft (in potential flying hours left, equipment, airframe status, etc.) were the eleven F1EDAs and two F1DDAs from the Qatar Emiri Air Force.

All these aircraft had been delivered to the QEAF between March 1983 and July 1984, except for QA63, delivered in 1992. Its operator was number 7 Air Superiority Squadron (call-sign "Falcon"). Of its original strength of twelve, two of the squadron's Mirages were written-off — QA72 "B" in 1991 and QA62 "T" in 1987 (replaced by QA63). During 1997 this unit will receive at least 12 Mirage 2000-5s, and possibly 15, including three two-seat conversion trainers; some of its pilots and engineers are already training in France. The complete package, including MICA missiles, Mic II, Super 530D, etc. is worth about $1.3 billion, less the $130 million trade-in value of the F1s.

Based at the military side of Doha International Airport, the Mirages participated, from 25 January 1991, in several strike operations during the Gulf War. They were always escorted by USAFE's F-16Cs from the 401st TFW, attacking Iraqi targets in the southern and central parts of occupied Kuwait. For this tasking they carried two 1,200 liter drop tanks under the wings, two Magic missiles on the wingtip stations, flare/chaff dispensers, and Remora ECM pod, as well as four 250-kg bombs on the centerline CLB4 bomb racks.

Their main task during that conflict was either to fly local air defense sorties or to stay in strip alert conditions. Those types of missions were flown with two Matra Super 530F-1 radar guided missiles, two Magic II IR-guided AAMs and full munitions for the 30 mm cannon. Also, during CAPs, 1,200 liter centerline tanks were carried. On the other hand, several reconnaissance missions were undertaken over Kuwait. Here they carried the COR-2 recce pod, as well as the Remora ECM pod and the conformal Sycomore flare/chaff dispensers. During Desert Shield/Storm,

A total of 24 Mirage IIIEEs (C.11 in the Spanish designation system) operated with Escuadrones 111 and 112 of Ala 11 (11th Wing) at Manises-Valencia Air Base, Spain. In fact Ala 11 celebrated its 40th Anniversary on September 6, 1995, having logged since then more than 250.000 flight hours in F-86F Sabres, T-33A Shooting Stars, Mirage IIIs and Mirage F1s. (author)

Seven two-stick Mirage IIIDEs (CE.11) were delivered to Ala 11, being shared by both squadrons, although 112 Escuadrón was disbanded on December 31, 1989. (Labajo)

Old and new, two 111 Squadron Mirage IIIs on the holding point besides the runway in company with a Grupo 15, Spanish AF, F-18B+ Hornet before a DACM mission. (author)

Doha International, the home of No. 7 Air Superiority Squadron, was also used by USAF and Canadian fighters.

Also evaluated by the Spanish Air Force were the Kuwaiti Mirage F1CK-1/2s, seven of the former and six of the latter. While there were no two seaters available because the four F1BK-1/2 were taken by the Iraqis and not returned; on the other hand, technical records had been destroyed and its operational capabilities were not as desirable, while the price asked for them was too high for the available budget. Also visited was Jordan, which offered its 13 surviving Mirage F1CJs. Again the price was too high and the offer was turned down. Finally, eight Iraqi Air Force Mirages embargoed in France were examined, but they were soon

With Captain Jesús García Labajo, a Mirage IIIEE takes-off in full afterburner for a night Combat Air Patrol during some air defence exercises over the Western Mediterranean in September 1992. (author)

The Mirage IIIEE/DEs were phased out of service of 111 Squadron on October 1, 1992, as an ambitious upgrade program had been canceled about one year earlier due to budget cuts. As nobody in the High Staff wanted to deactivate this unit —the first Spanish AF jet fighter outfit—, the temporary solution was to take eight Mirage F1CE and or F1EE (C.14A and C.14B in the Spanish designation system) on loan from Ala 14 (14th Wing), which is the Spanish AF main F1 operating unit; Ala 14 is based at Los Llanos-Albacete Air Base. This transition also took place on October 1, 1992. (author)

Eight of the initial jocks of the new 111 Squadron, on October 23, 1992. With an established strength of 13 pilots, six were from Mirage IIIs converting to the F1 at Albacete, while seven came directly from the F1 squadron at Gando air base in the Canary Islands, the loaned F1s maintained their Ala 14 fuselage codes. (author)

Ala 11 insignia.

discarded, as they comprised four two seat F1BQs and six single-seat F1EQ-7s, too many of the former and too few of the latter. On the other hand, France offered four F1Cs and one F1B as exchange for two CASA C-235 transports, and this offer was taken up. The five Mirages were delivered to Ala 14 from November 1994, one per month, until March 1995.

Early in 1994 the three sided negotiations started in earnest — France, Qatar and Spain — for the acquisition of the 13 Mirages. The French part of the deal was very important, as there was a multimillion dollar contract with the Qataris, not only involving the 12 Mirage 2000s, a large stock of spares, weapons and training, but also the construction of a new air base from the ground up in order that the QEAF should operate from a proper airfield, and be freed from the restrictions imposed by operating from a civilian airport as is Doha International. During the negotiation, it was agreed that Dassault would provide technical support for the ex-Qatari F1s delivered to Spain, supplying maintenance services and detaching a technical representative to Manises, etc. These negotiations continued almost to the eve of the aircraft's departure for Spain. The price was 17,500 million pesetas for the 13 aircraft, an advanced flight simulator, air-to-air missiles, and a complete stock of spares, to be paid during the four years of the program (1994/1997). As said the potential flight hours of F1EDA/DDAs are quite good, and the Mirage could remain in service until the year 2010/2013. In QEAF service they have flown relatively few hours, and their average life since delivery is ten/eleven years, while the last two-seater is one of the last F1s built and has only three years left. (See Box #1 for full serial listing, both Qatari and Spanish).

The signed contract specified that seven should be delivered during the Summer of 1994, and the remaining six during the second half of 1997, An important stock of spares, the flight simulator, maintenance equipment and both captive and 40 operational Matra Super 530F-1 missiles should arrived in Valencia as ship's cargo before the end of 1994, including about half of the seven spare Atar 9K-50 engines.

The Mirage F1EDA/DDA is almost a second-generation Mirage F1, the single seat has in fact the pre-installation for adding

Flown by the late Christian de Haya Beyrer, Mirage F1CE 14-10 assigned to 111 Squadron landing on runway 30 at Manises-Valencia on September 14, 1993. (author)

Captains Jesús Garcia Labajo (foreground) and Agustin Serrano (background) sign the aircraft's books before a mission. (author)

Blue Mirage F1EE with Ala 14 codes, but flown by an 462 Squadron takes fuel from a Boeing 707TT from 45th Group. (author)

One of three F1BEs and a single F1B twin stickers in Ala 14 service, which mostly are assigned to 142 Squadron which has the Operational Transition Unit for the Spanish F1s (besides the first line tasking of AWX (all-weather Interceptor) and FBA (fighter-bomber attack), like the other Mirage squadrons. (author)

Composite F1EE (foreground) and F1CE (background) formation of 141 Squadron, November 1994. The F1EE carries under the left wing a Barax jamming pod, while on the vertical fin carries the new spanish developed AN/ALR-300 Radar Warwing Receivers, noticeable from the French BF system, by the square antennas on the leading and trailing edges, this system, similar to the AN/ALR-67 installed on F/A18C/D Hornets is being retrofitted to the whole fleet. (Author)

111 Escuadron insignia.

the air-to-air refuelling probe, including the 8 cm plug behind the radome. The most noticeable thing about these fighters is their desert camouflage scheme, brown and beige on the upper surfaces and mid-blue on the undersurfaces. But these differences will be better felt by 111 Escuadrón pilots.

The C.14C/CE.14C (their official Ejército del Aire <Spanish AF> designation), have an integrated nav/attack system, associated to a Dassault Electronique 182 digital mission computer, which receives and processes all the information provided by other sensors: Cyrano IV radar (with four air-to-ground modes), Sagem Uliss 47 INS, etc. Most of this information is projected in a Thomson-CSF VE-120 HUD, while the armament panel is also different from the Spanish variants, as well as some changes in the cockpit instrumentation, including SEM Martin Baker F10M, capable of zero-zero operations.

The Qatar Emiri Air Force insignia.

On August 23, 1994 four F1EDAs (C.14C) and one F1EDA (CE.14) arrived to Manises-Valencia AB after a two-day ferry flight after stops at Jeddah (Saudi Arabia); El Berigat AB (Egypt); Tanagra AB (Greece) and Solenzara AB (Corsica); for the flight they carried two 1200-liter drop tanks. (Author)

Due to the greater capability of this variant in air-to-ground operations, 111 Escuadrón has changed its mix of roles. Previously it only had F1CEs, that mix was 60% air-to-air and 40% air-to-ground; now 60% is devoted to the former (Air Interdiction <AI>, Offensive Air Support <OAS>, etc.), while as an All Weather Fighter it will provide the other 40% in OCA/DCA (Offensive Counter Air/ Defensive Counter Air) missions. If the medium range radar-guided Matra Super 530F-1s have been acquired, the short-range IR-guided Magic 2s have been rejected, as all the Spanish AF Mirage F1s are equipped with the AIM-9JULI Sidewinder (a modernized AIM-9N/P with the AIM-9L seeker, which can be launched head-on and from the sides with a high kill probability).

The Qatari F1s also have other equipment which will be acquired in the near future, perhaps when the second batch is delivered or even before. These comprise the Dassault COR-2 recce pod, Thomson-CSF Remora ECM pod and Matra Sycomor flare/ chaff dispensers. Also these Mirages are wired for operating the Matra ARMAT antiradiation missiles, as well as the TMV 018 Syrel electronic intelligence (ELINT) pod — already operated by Spanish F1EEs.

It is also possible that a few more of the excellent Dassault Electronique Barax ECM pods will be acquired to equip them and, when they reach their time for the first general overhaul (Grand Visite 1) at the Albacete's Air Depot (3,000 flight hours or 13 years'

In 19/94 the Spanish Air Force acquired the 13 surviving Qatar Emiri Air Force (QEAF) Mirage F1s (eleven F1EDAs and two F1DDAs) through the good offices of Dassault which also had signed a contract with Qatar for the delivery in 1997 of 12 multirole Mirage 2000-5s. These fighters, in very good condition, were earmarked for service with 111 Squadron. Five were delivered on August 23, 1994, two on September 1, and the remaining six during the last quarter of 1997, or earlier, allowing the return of four F1CEs to Ala 14. Illustrated is one of the Mirage F1EDAs at Doha air base, with the Qatari roundel being replaced by the Spanish roundel. (QEAF)

A Qatari officer salutes Captain Jorge Albalat of 111 Squadron at Doha, mid-August 1994, in the background are Captains Bas and Labajo. (QEAF)

life), they will get their BF radar warning receiver replaced by the Spanish designed AN/ALR-300. The desert scheme will change to the standard overall light grey of the rest of the Mirage fleet.

Ala 11 detachment at Qatar was formed by 48 men; the first departing for Doha on 18 July 1994, followed by a second on August 1st. All groundcrews (under the command of engineering officer Major Felipe Ramajo Romero), the eight designated pilots (all from 111 Escuadrón, including its C.O., Major Felix Sahagún Schwartz), six captains (Jorge Albalat, Jesus G. Labajo, Jesus Elices, Javier Bas, Jose Miguel Quintana and Sergio Riego) including two reserves (under the command of Lt. Col. Carlos Sebastiá Camarasa — Ala 11 Operations Group C.O.), departed from Manises on August 8th. The three groups made their flight in Grupo 31 C-130H Hercules, as this unit had a meritorious part in the operation.

During their stay at Doha, each pilot logged two flights, one in F1DDA and another in DDA, plus six intensive simulator sessions. Once the last difficulties were sorted out, the Mirages became Spanish property on August 12. On Sunday, August 21st they started the ferry flight to Manises. The formation was composed of four F1EDAs and one F1DDA, which took off from Doha's 15,000 foot long runway at 07:30 hours local time, with the call sign AME-11. They were followed by a C-130H with groundcrews, spares, etc., while previously, on Saturday, the other Hercules had departed heading for the first stage.

Thus, during all the ferrying operation, one of the Hercules preceded the formation and the other trailed it, providing support and SAR cover in case of any difficulty. However, this long ferry operation developed as planned, and there was no problem in the approximately 4,000 nautical miles separating the runways of

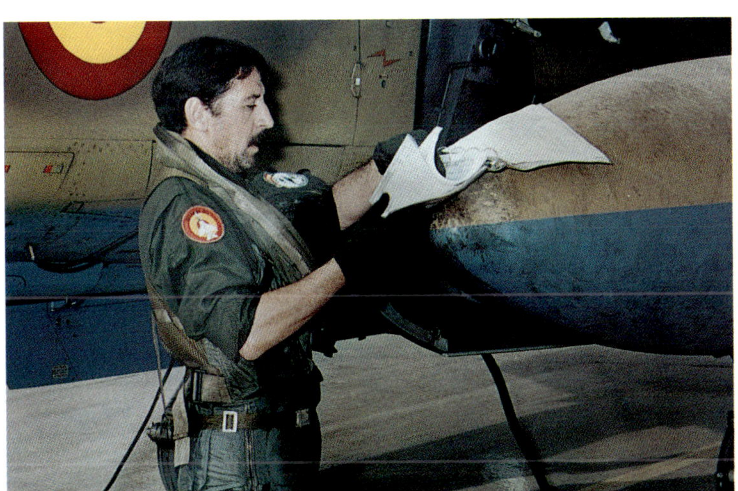

Lt. Col. Carlos Sebastia Camarasa, the Commanding Officer of Ala 11 Operations Group leaning the aircraft's book against one of the RP-35 1200-liter drop tanks just before staring the ferry flight from Qatar to Spain. (QEAF)

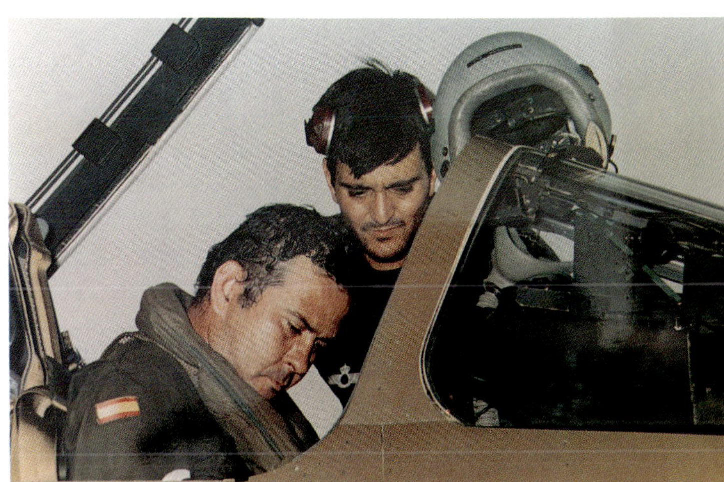

Major Felix Sahagun, 111 Squadron C.O. from October 1, 1992 to September 1995, performing the last cockpit checks before the ferry flight from Doha to Manises-Valencia, on August 21, 1994. (QEAF)

This picture of 111 Squadron pilot Captain Francisco "Paco" Almenta shows typical Spanish Af jet pilot gear, including the American HGU-55/P helmet and the British Mk.30S survival vest. (author)

Doha and Manises. This amounted to almost nine flying hours, in four stages and two nights.

The first stage was from Doha to Jeddah International Airport, on the shores of the Read Sea, near Meca, in Saudi Arabia. Aufter refuelling, they flew to El Berigat Air Base in Egypt, staying overnight. Next morning the next leg brought them to Tanagra AB in Greece. After a quick lunch, they took off for Solenzara air base in Corsica, where they spent the night. On the morning of August 23rd, they launched again, heading for Manises air base, where they arrived at about 11:30 hours local time. The aircraft involved were: F1EDA: QA71/A; QA73/C; QA74/D; QA75/E. F1DDA: QA62/U, all carrying their Qatari serials and Spanish roundels painted over their former QEAF roundels. On September 1st two 111 Escuadrón pilots (Jesus Elices and Sergio Riego) ferried F1EDAs QA73 and QA74 to Dassault's overhaul/maintenance shops at Bordeaux, for a "P" Inspection (600 flight hours or 30 months). On the same day, they brought to Manises aircraft QA75/F and QA79/I, which had just finished a similar inspection, while QA73 and QA74 were returned to 111 Escuadrón in November and December 1994 respectively. By that date, follow-on second echelon maintenance will be performed by Ala 11 Maintenance Squadron. On the other hand, 111 will have, in strength, at least

Photographed almost a year after their arrival to Spain, the 111 Sqn. only twin-stick F1DDA (the other, together with five single sticks will arrive in 1997) with the roundel applied at the correct size and position, while the Ala 11 insignia has taken its place. (Author)

until 31 December 1997, four Mirage F1CEs loaned by Ala 14, while the other four were already returned in October 1994.

For most of 1995 111 Escuadrón had an authorized strength of 15 pilots, 12 posted to the squadron itself, including the C.O. (Maj. Félix Sahagún - to be replaced by Major Rafael Nuñez in due course). The remaining eleven are Captains or "indians" as this rank in known in Spanish AF fighter pilot slang. Each jock gets some 150/180 flying hours per year as an average, the unit being divided in two flights (1a Escuadrilla and 2a Escuadrilla - 1st Flight/2nd Flight). The other three pilots are Ala 11 staff officers attached to the squadron, including the wing C.O. Thus, if the Mirage III era at Manises lasted 22 years, it is quite possible that the same will happen with its second-generation stable mate, the sleek Mirage F1. In fact, according to plans, 111 Escuadrón will be the last to dispose of them and reequip with the controversial Eurofighter EF2000 . . .

111 Squadron flight line at Manises-Valencia, August 1995, in the foreground an F1EDA, in the background one of the four F1CEs loaned by Ala 14 (which will be returned when the six remaining Qatari Mirages arrive to Manises-Valencia). The F1EDA is a better multirole platform than the F1CE or F1EE, as its nav/attack system is fully integrated, while its Cyrano IV radar has four air-to-surface modes, providing data to the VE-120 HUD thus the pilot can select automatic CCIP (Continuous Computed Impact Point) CCRP (Continuous Computed Release Point) bombing modes, with a high degree of exactitude. (author)

Close-up of a Mirage F1EE flying with 141 Squadron, on the intake wall it carries Ala 14 insignia. The Spanish Air Force is to undertake an upgrade program for the Mirage fleet, involving 55 aircraft which will last until 1999, and will amount to install an integrated Nav/attack system centered around a MIL-STD 1553B data bus, a computerized mission planning system/debriefing system using a removable cartridge, as well as a new cockpit layout, including a digital multipurpose display and an holographic HUD, as well as a ring-laser INS integrated with a GPS, expanding the aircraft's ground attack capabilities. (author)

Major Mariano Diez, C.O. 462 Squadron in 1995. (author)

Pictured in April 1993, this Mirage F1EE wears Ala 14 codes, although it is assigned to 462 Squadron, note that all Spanish F1s (except the Qataris) are being painted in an attractive light grey scheme known as LIR (low-infrared) polyurethane supplanting the earlier lizard and light blue schemes, also painted is a mock cockpit below the original. (author)

SPANISH SERIALS OF THE EX-QEAF MIRAGE F1EDA/DDAs ASSIGNED TO 111 SQUADRON

MIRAGE F1EDA

QEAF	DASSAULT	C/N	SPANISH AF	SERIAL	DELIVERY
QA71	"A"	510	C.14C-74	"11-01"	23 Aug. 94
QA73	"C"	552	C.14C-75	"11-02"	23 Aug. 94
QA74	"D"	556	C.14C-76	"11-03"	23 Aug. 94
QA75	"E"	560	C.14C-77	"11-04"	23 Aug. 94
QA76	"F"	567	C.14C-78	"11-05"	1 Sept. 94
QA77	"G"	572	C.14C-79	"11-06"	— 1997 —
QA78	"H"	577	C.14C-80	"11-07"	— 1997 —
QA79	"I"	580	C.14C-81	"11-08"	1 Sept. 94
QA80	"J"	588	C.14C-82	"11-09"	— 1997 —
QA81	"K"	593	C.14C-83	"11-10"	— 1997 —
QA82	"L"	594	C.14C-84	"11-11"	— 1997 —
MIRAGE F1DDA					
QA62	"U"	539	CE.14C-85	"11-70"	23 Aug. 94
QA63	"V"	704	CE.14C-86	"11-71"	— 1997 —

In the Spanish AF designation system, the first C designates the type, in this case Caza (Fighter), the number 14 is the fourteenth type of fighter operated by the Spanish AF since its formation in 1939, and the second C means the third variant of the Mirage F1 in Spanish service. Therefore, the F1CE is the C.14A and the F1EE is the C.14B, while CE means Caza Enseñanza (Fighter Trainer). The serial is applied on both sides of the vertical fin, while the fuselage number, usually below the cockpit, is the number assigned to the unit which operates the aircraft: thus, 11 - stands for Ala 11 (11 Wing). On the other hand, all other Mirage F1s (CE/C/EE/B/BEs) have Ala 14 fuselage numbers, as this unit is the centralized operator of these variants — although some are loaned to 111 Sqn. at Manises and 462 Sqn. at the Canary Islands, but without changing the side numbers.

111 Squadron two-sticker 11-70 photographed on October 19, 1994, during a practice intercept sortie against to Ala (12th Wing F-18 Hornets). The front cockpit was occupied by Captain Rafael Hernandez Micó, a recent arrival to the squadron which made part of the F1 course in France at Reims air base, while the remaining of the syllabus was made with the squadron, rear seat is occupied by Captain Jorge Albalat Estela, the squadron's instructor pilot. Note that the aircraft still has the roundel placed in the wrong place. (author)

The impact of two GBU-10 Paveway II laser guided bombs dropped by two Mirage F1s from 111 Sqn. at the live weapons training range of Las Bardenas Reales, northern Spain, which until 1992 was also used by the USAFE. (author)

With Captain Jesus Gomez Ruiz at the controls Mirage F1EDA "11-03" from 111 Squadron lands on Runway 30 at Manises-Valencia air base, in spite of its relative small size, the Mirage F1 holds in its internal tanks a considerable fuel quantity, 4100-liters. (author)

Besides 111 Sqn., there are three other fighter squadrons in the Spanish Air Force flying Mirage F1s, these are 141 Squadron, whose badge is illustrated in this picture and which was the first Spanish F1 unit, activating on June 18, 1975, when the first four F1CEs arrived to Albacete. (author)

Still without unit codes and insignia, this Mirage F1EDA (still with the RAF style code letter applied by the QEAF on both sides of the vertical fin) was ferried from Dassault overhaul/repair plant at Bordeaux, to Manises-Valencia, by 111 Sqn. pilot Sergio del Riego on September 1, 1994. (author)

142 Squadron, also of Ala 14 was formed in April 1980. (author)

Captain Jose Miguel Diaz Quintana, was posted to 111 Squadron where he stayed until December 1994, flying Mirage IIIs and F1s, then in January 1995 transferred to 141 Squadron of Ala 14, based at Los Llanos-Albacete where he continues to fly Mirage F1s. (author)

462 Squadron, based at Gando air base, Canary Islands, was formed in 1982, by renumbering 464 Squadron which then was equipped with Northrop F/RF-5As. (author)

Since 1978 Ala 14 maintains a readiness flight formed by two Mirage F1s, in five minutes during daylight and fifteen minutes at nightime, both Mirages are armed with two AIM-9N3 Sidewinder missiles and full ammunition for its 30mm guns. (author)

Flight suit shoulder patch carried by 462 Squadron pilots. Actually this unit (call-sign "Hawk") has on strength(?) eight Mirages, five F1EEs and three F1CEs (or F1Cs). (author)

Mirage F1CE flown by a 141 Squadron pilot in landing configuration. The Spanish Air Force acquired in total 45 Mirage F1CEs, 22 Mirage F1EEs, six two-stick F1BEs,in the period 1975 to 1982, of which 30, 17 and three remained in service by late 1995, additionally, besides the Qatari examples related elsewhere, four F1C and one F1B from French Air Force (Armée de l'Air) stocks were acquired in 1994 for Ala 14. Ala 14 is the main operating location for the Spanish fleet of F1s (except the Qatari examples) and thus actually all the aircraft carry the fuselage codes of this wing, even if they are serving with 462 or 111 Sqn.s. (author)

Photographed in October 1989, still in their unit codes, two Mirage F1EEs take a drink from an KC-130H of 31st Transport/Tanker Group. The F1EEs (C.14B in Spain) have some differences in respect of the F1CEs, the most noteworthy is a fixed air-to-air refuelling probe and an inertial navigation systems, which permits these aircraft to operate the Thomson-CSF Syrel ELINT pod -which actually is one of the tasks assigned to 141 Squadron at Albacete. (author)

Mirage F1CE "14-43" while in service with 111 Squadron in January 1995, the aircraft carries a RP-35 1.200-liter drop tank, a Barax ECM pod and two AIM-9N3 Sidewinders. Spanish F1s have the improved Thomson-CSF Cyrano IVM radars. (author)

With airbrakes deployed, a Mirage F1CE assigned to 462 Squadron in flight over the Atlantic Ocean. (author)

One of four Mirage F1Cs received (December 1994, January, February and March 1995) in Ala 14 from French Air Force stocks, sits in front a Hardened Aircraft Shelter (HAS) at Los Llanos-Albacete air base. (Author)

Mirage F1CE "14-16" photographed at Los Llanos-Albacete in March 1995 while in service with 462 Squadron, on the centerline pylon it carries a SUU-20 practice bomb dispenser.(Author)

Initially, all the 22 Mirage F1EEs were acquired for 462 Squadron at Gando (part of 46th Composite Wing, Spanish AF), and thus wore Ala 46 codes, but towards the early nineties, as Ala 14 became the head of F1 operations, the squadrons started to operate with mixed equipment EE/CEs, and thus 462 lost the codes. This picture shows F1EE "46-16" in clean configuration. (via Javier Bas)

111 Squadron, Spanish Air force, pilots in February 1996, the author is fourth from the left on the standing lane. (author)

View forward from the rear cockpit of a 111 Sqn. Mirage F1DDA two-seater flying over the Mediterranean. (author)

Composite formation of 111 Sqn. Mirages, including two F1EDAs (desert camo), two F1CEs (grey camo), and an F1C (blue camo). (author)

Two Mirage F1EDA in welded wing formation during a hard right turn. (author)

A Mirage F1EDA of 111 Sqn. flown by Capt. Jesus Garcia Labajo flies low over the Mediterranean, the fighter is loaded with two Matra Super 530F1 medium-range radar guided AAMs. (author)

A bomb carrier, with small practice bombs, beside a 111 Sqn. Mirage F1EDA, at Manises-Valencia air base, March 1996. (authot)

A Spanish Mirage F1EDA seen in landing configuration over the Mediterranean sea. (author)

Armorers loading the DEFA 553A 30mm guns of a Mirage F1EDA.

The Thomson-CSF VE-120 HUD as fitted to the Mirage F1EDA in the navigation mode. (author)

Cockpit of the Mirage F1EDA. (author)

The Matra Super 530F1 is a radar guided medium range AAM, with a maximum range of about 35 Km, being capable of intercepting targets at up to 75, 000 ft. This missile was heavily used by Iraqi F1s during the war against Iran. (author)

Mirage F1EDA loaded with six 250 Kg bombs. (author)

Mirage F1EDA of 111 Sqn. overflying the city of Valencia. (author)

Mirage F1EDA of 111 Sqn. with two Matra Super 530F1 AAMs, note that the missiles are painted in the same color as the lower surfaces of the Mirage. (author)

Two Mirage F1EDAs of 111 Squadron, home based at Manises-Valencia, caught during a training hop on 22 February 1996, the aircraft are flown by Captain Labajo (foreground) and Capt. Bas (background). (author)

Taking off from runway 30 at Manises-Valencia air base is Captain Jesus Garcia Labajo in "11-04". (author)

11

Other Battles

AEGEAN TOP COVER

In the mid-seventies a new shape appeared over the deep blue sea that separates two NATO allies but also two long time adversaries, it was the Dassault Mirage F1CG, of which the Polemiki Aeroporia acquired forty. Today, complemented with Mirage 2000s and F-16 Fighting Falcons, the surviving F1s continue to provide a valuable asset for the Taktiki Aeroporikis Dynamis (Tactical Air Force), equipping two all-weather fighter squadrons, at Tanagra and Iraklion respectively. The author interviewed the pilots and the C.O. of 342 Mira, providing some insight into its operations and combat role.

The first Armée de l'Air Mirages to be based in African territory, where the Mirage IIICs from Escadron de Chasse 3/10 "Vexin" —EC 3/10-(Fighter Squadron), activated in April 1978 at Creil air base France, transferring in flight to Djibouti air base in the Republic of the same name, on Eastern Africa, ten years later it was replaced by the Mirage F1Cs and renumbered as EC 4/30, and in 1995 it was again renumbered as EC 4/33. (Author collection).

Although belonging to the now defunct EC 1/30 Valois, this Mirage F1C camouflage scheme denotes that it served with then EC 4/30. Actually the number plane EC 1/30 exits for a F1CT squadron at Colmar air base.

For many years the Armée de l'Air has deployed a substantial combat element at the Republic of Chad in Africa, to stop the territorial ambitions of Libya, which claimed an oil rich territory north of the country. Illustrated is a Mirage F1CR from Escadre de Reconnaissance 33 on patrol near the disputed territory. (Sirpa Air)

One of the most important French types during the Gulf War was the Mirage F1CR, which with their accurate nav/attack system made all weather pin-point attacks against Iraqi targets, Escadre de Reconnaissance 33 pilots performed many hundreds of hours in combat sortie without any loss, both in recce taskings and in the classic bomber routine. (Sirpa Air)

Also from December 1990 to March 1991, Escadre de Chasse 12 provided an intercept detachment—twelve Mirage F1C-200—at Doha air base, Qatar, to protect the oil-rich emirate, many CAP sorties were flown, but no Iraqi aircraft were encountered. (Sirpa Air)

Two Armée de l'Air pilots, one wears the desert type flight suit, issued to aircrews during the Gulf conflict. (Author)

In 1974 Greece withdrew from NATO's military structure in response to the Turkish invasion of Cyprus (she returned in October 1980). Its Polemiki Aeroporia (Military Aviation) was starting to make advanced plans to upgrade its air defense capabilities, which mostly rested on F-104G Starfighters and old F-102A Delta Daggers provided by the USAF in 1969. Thus, the confrontation with Turkey added a new priority to this selection, which also involved the acquisition of F-4Es (mostly dedicated to the fighter bomber role), A-7Hs and T-2Cs. The Hellenic Air Force needed a modern, but not too sophisticated, interceptor and Dassault provided the right aircraft at the right time — the then new Mirage F1.

In June 1974 the Greek government signed a contract for the acquisition of forty aircraft. In order to hasten the deliveries, the first eleven aircraft were diverted from the Armée de l'Air production line. Seven months after signing the contract, a nucleus of Greek pilots were sent for training to France. First they went to Mont-de-Marsan and later to Reims, allowing the ferrying of the first two Mirage F1CG — which had been handed out to the Greeks at Mont-de-Marsan in February 1975 — to their base of Tanagra in August of that year, fourteen months after the order had been placed. The 40th, and last, Mirage was delivered in April 1978.

After twenty years of operations, some 32 Mirage F1CGs remain in service. Each airframe has logged, on average, some 4,000 flight hours, having passed two general overhauls (Grand Visite) which are performed by Hellenic Aerospace Corporation. Their overall condition is good, although there are the usual problems with aircraft which have logged so many flying hours. Still, the HAF plans to keep them in operational service for at least another ten years, and it has planned to execute a limited upgrade of the fleet, especially the nav/attack system, including the addition of flare/chaff dispensers.

The Mirage was the first non-American jet fighter incorporated into HAF service. The HAF had to get used to the peculiar French supply system, standards and high priced spares (which implied that sometimes the operational availability of the fleet was not as high as desired).

Although classed as all-weather interceptors, Greek Mirage F1CG have only short range IR-guided AAMs, albeit the all-aspect capable AIM-9P4 Sidewinder, of which they carry four in a

The Mirage 2000Cs from Escadre de Chasse 5 were the main French Air Force fighters during Desert Shield and Desert Storm, based at Al Asha, Saudi Arabia, made countless CAP and escort missions over Iraq, but never encountered an IrAF fighter, it seems all the trade was for the USAF's F-15s. (Sirpa Air)

The latest Mirage F1CT has been booted in combat during Deny Flight operations over Bosnia, especially since Operation Deliberate Force started in earnest on August 30, 1995, a detachment is maintained at Istrana air base, Italy, from their home base of Colmar, France. (Author)

unique and exclusive installation. Besides the two on wingtip launchers, another two are carried on the underwing outer pylons on Aero 3B launchers, duplicating the F1's meager capability in air-to-air smart weaponry. Besides, they have the two highly effective DEFA 553 30mm cannons, each with 135 rounds.

THE SQUADRONS

The first F1 operational unit was 342 Mira Pantos Kerou "Sparta" (All-Weather Fighter Squadron), based at Tanagra, some 65 km northwest of Athens. It now forms part of 114 Pterix Mahis (Combat Wing) with two Mirage 2000C/B squadrons, 331 Mira "Aegeas" and 332 Mira "Geraki." 342 Mira was established in 1955 as a Day Interceptor Squadron operating F-86E aircraft and had that role until 1964 when the Sabres were supplanted by F-84F Thunderstreaks, operating in the fighter bomber role until 1969. In 1969 342 Mira was reequipped with the F-102A/TF-102A and took up an All-Weather Interception (AWX) role, which continues to this day. In 1975 "Sparta" started conversion to Mirage F1CGs. The first two aircraft of this type landed in Tanagra Air Base in August 1975. By May 1976 the squadron had completed conversion and training of pilots and technicians and was declared operational.

Actually, 342 Mira is led by 41-year old Lt. Col. Alexandros Papagermanos. Papagermanos has some 2,100 hours in his logbook, including 400 on F-5s and 800 in F1s. The squadron is divided into three flights, two operational and one tasked with the conversion of young pilots to the Mirage F1CG — before being posted to Tanagra (or any other HAF unit). Recently graduated pilots had performed some 215 flight hours during three years in the primary T-41 (35 hours), basic T-37 (80 hours) and advanced T-2C, including weapons training (100 hours). In total, 342 Mira has fifteen pilots assigned, the Lt. Col., a Major, six Captains, six 1st Lieutenants and one 2nd Lieutenant. To acquire the CR (Combat Ready) Status, a young pilot has to perform two F1 training syllabus for about one year, including theoretical training, 25 simulator hours (the only HAF F1 simulator is at Tanagra), plus 60 flight hours of basic handling, continuing with another sixty hours learning to use the F1 to its maximum as an interceptor and a fighter. When there are training commitments, 342 Squadron loans a few aircraft from 334 Squadron which has a higher authorized strength. After that, the pilot can perform most of the squadron's missions, including the vital QRA mission — which 342 Mira shares each week with the two Mirage 2000 squadrons — in which two aircraft are in five-minute readiness 24-hours a day.

Left: Mirage F1CRs from Escadrons de Chasse 1/33 and 2/33 forward deployed at Istrana, provide a valuable reconnaissance capability for Operation Deny Flight. (Author)

The Indian Air Force received in the period 1985-1988 a total of 46 Mirage 2000, comprising 42 Mirage 2000H single seaters and four Mirage 2000TH two-seaters, serving with No 1 "The Tigers" Squadron and No 7 "Battle Axe" Squadron based Gwalior Air Base, these fighters, with ground attack capabilities perform numerous patrol missions over the disputed Kashmir region, near the border of Pakistan. First picture shows Mirage 2000H KF123 from No 7 Squadron and the second KF140 from No 1 Squadron at its base. (Steinemann)

Jordan's RJAF received a total of 17 Mirage F1CJs and two F1BJ paid by Saudi Arabia, serving with their primary mission of interceptors with No 25 Squadron at Azraq Air Base, these are painted in a two-tone light grey scheme. (Steinemann)

These were followed by another batch of seventeen, although these were multirole Mirage F1EJs serving with No 1 Squadron at the same air base, their noteworthy feature is their lizard camouflage scheme. Due to the volatile climate in the Mid-East these Mirage have always been in a high state of readiness, its pilots are among the best of any Arab air force. (Steinemann)

The FAE (Ecuatorian Air Force) operates the survivors of its 16 Mirage F1JA and two Mirage F1JE trainers with Escuadrón 2112 of Grupo 211 home based at Taura air base. These fighters shot down two Peruvian Su-22 Fitters in air combat on 10 February 1995 during the border war of January/February between the two countries. (Steinemann)

This picture shows a representation of the three combat squadrons of Grupo 211 of the Ecuatorian AF, Escuadrón 2111 equipped with Jaguars, 2112 with Mirage F1s and 2113 with Kfirs, forming a small but well balanced and trained fighting force, complemented by A-37B and AT-33 light attack squadrons. (Steinemann).

In all Peru received a total of 40 Mirage 5Ps and 5DPs, of which ten were sold to Argentina in June 1982, actually fifteen 5Ps and three 5DPs remain in service with Escuadrón 611 at Chiclayo air base, being heavily involved in the war against Ecuador in January/February 1995, although no losses were reported. Also the FAP (Fuerza Aerea del Peru) received in 1985/1986 ten Mirage 2000P and two Mirage 2000DP, operator unit is Escuadrón 412 ar Mariano Melgar air base, near La Joya, Peru. (Dassault).

Colombian Kfir C7s, operated by Escuadrón 213 (the upgraded Mirage 5COA/COR and COD serve with Escuadrón 212) have been used in support of the army in anti-guerrilla and anti-drug traffickers operations, even mounting CAP operations to intercept aircraft used by the drug lords to transport it up north. (Steinemann)

If the pilot participates in any scramble, according to his experience, he is limited in the minimum engagement altitude — to 8,000 ft. for the less experienced, passing through to 5,000 ft., and at very low levels for the most experienced. Each pilot, once posted in any of the two operational flights, records some 13 flight hours per month. 342 Mira has an authorized strength of 12 Mirage F1CGs. Interestingly, the Mirages of 342 Squadron sport the names of different Greek islands under the right side of their cockpits. A total of 18 aircraft have received them which, for the record, are: "Psara," "Naxos," "Santorini," "Kos," "Kastelorizo," "Ikaria," "Samos," "Limnos," "Folegandros," "Samothraki," "Tinos," "Skiros," "Kea," "Karpathos," "Thasos," "Lesvos" and "Kriti."

342 Mira's role, as mentioned previously, is AWX. The average capability of their pilots is quite high, mainly the most experienced. They are very aggressive in air combat and have sound tactics, knowing their mount perfectly and how to operate it to its limits in order to compensate for more modern equipment in the hands of the adversary like the F-16. DACM is carried out frequently with the Fighting Falcons of 111 Pterix Mahis, as well as with Phantoms, Freedom Fighters and the collocated Mirage 2000s. Although Tanagra is the main operating base of 342 Sqn., usually it sends detachments to advanced air bases located at three Aegena islands, Santorini, Limnos and Skiros. This considerably reduces the scramble time when any Turkish aircraft violates its airspace. This would not be possible without the hard work performed by the ground crews, capable of maintaining in tip-top condition aircraft which are rather old and which have seen heavy usage.

334 Mira Pantos Kerou "Talos," part of 126 Sminarkia Mahis (composite wing) is the other HAF Mirage F1CG squadron. Since the summer of 1989, "Talos" has been posted from Tanagra to Iraklion air base on the island of Crete. From there the unit, led by Lt. Col. George Skimpas, and with an authorized inventory of 16 Mirages, covers in the AWX role for the southern Aegean and its approaches. "Talos" also has two operational and one training flight, the latter with the same tasking as that of 342 Sqn., except that the young pilot had to go to Tanagra for simulator training. 334 Sqn. also maintains two F1s on QRA, but in this instance this is performed the year round.

DOGFIGHT OVER THE AEGEAN

Encounters of Hellenic and Turkish fighters have taken place almost since both countries became NATO allies. According to Greek authorities, the two have been at odds since the Turkish invasion of Cyprus in July 1974. TAF fighters have repeatedly violated Greek airspace, endangering commercial flights in the area covered by Athens FIR. HAF interceptors quickly scramble to challenge the intruders and these interceptions usually develop into simulated air combat engagements, often at very low levels over the Aegean.

IAI Kfirs from Escuadrón 2113 were also heavily involved in combat during the early 1995 war with Peru, one of them shot down a Peruvian A-37B Dragonfly of 10 February, while two Kfirs were hit by SAMs, possibly SA-14s, one made back to an advanced airfield, while the other crashed. (Steinemann)

During the early seventies Lybia, already under the mercurial leadership of Col Kaddafi, received 110 Mirage 5s, comprising 53 Mirage 5Ds, 15 Mirage 5DDs, 32 Mirage 5DEs (like the one illustrated, which is equipped with the Cyrano Ibis fire control radar) and ten Mirage 5DRs. The aircraft main base is Nasser AB, near Tobruk, and during their operational life they were involved in the border war with Egypt in 1977, in the Yom Kippur War against Israel in 1973 (based in Egypt, with several being lost) and in the war in support of Chadian rebels in the north of that country. (Cuny)

This is what happened on 18 June 1992 to Lieutenant Nicolaos Sialmos of 342 Mira. On that date he was on QRA with Captain Ioannis Manolakos, when the signal was received at Tanagra, from the Hellenic Tactical Air Force (HTAF) sector operations center sector at about 11:40 hours, to scramble the QRA pair. Lt. Sialmos (who was the wingman) was flying aircraft 116 nicknamed "Lesvos." Just after take-off and at about 5,000 ft., just upon reaching the sea, both fighters adopted combat spread tactical formation, turning on the armament switch which put the Sidewinders and guns on "live." Radar vectored them toward two bogeys in the southeast, near the island of Agios Efstratios, and at very low level. The sea had some mist, but at some four miles the intruders were detected almost head-on. At 12:00 hours Capt. Manolakos, the section leader, advised them through the guard channel that they were in Greek airspace and to abandon their approach.

Instead, the Turkish F-16s turned into the Mirages and Lt Sialmos F1 was locked by one of the Fighting Falcon's radar. As he called his leader that was what showed on his BF radar warning receiver's indicator, evasive action was taken and a dogfight developed at very low level. This was a very dangerous situation due to the mist. After several turns the combat turned into some kind of "sandwich." An F-16 had an F1 on its tail, which itself had another F-16 and finally an F1. Sialmos felt an overwhelming enthusiasm and this feeling overwhelmed himself. In order to get better results during his effort he flew lower, despite the warning of his leader; then the Fighting Falcon he was pursuing made a high-g level turn to evade from the persistent Greek pilot (had

Mirage 5D wearing Lybia's old insignia over its original three tone camouflage scheme. (Dassault)

The Libyan Arab Air Force was also a customer for the second generation Mirage, acquiring a total of 38 machines, comprising 16 Mirage F1EDs, 16 F1ADs and six F1Bds, serving with two squadrons at Okba ibn Nafa air base, near Tripoli. Detachments were made to Maaten as-Serra, southern Lybia, performing several attacks into Chad during the late eighties, until Armée de l'Air Mirage F1Cs started to mount CAP missions over the area, then Libyan air operations was reduced. (Dassault)

weapons been used in that engagement, it is quite sure that Lt. Sialmos could have shot down this fighter). This was a relatively easy maneuver for the F-16, but more difficult to perform in the F1 which had less thrust, and the Mirage crashed into the sea close to the island of Agios Efstratios. Quickly the TAF fighters disengaged and the HAF mounted a rescue effort, but to no avail. Lt. Sialmos had been killed instantly.

On 8 February 1995 the score evened. This time it was 334 Mira's turn to scramble. Mira sent two of the F1CGs to intercept a pair of Turkish F-16s near the island of Rhodes. The usual engagement took place, starting at 11:33; but, some two minutes later one of the F-16s had an engine flameout due to a fuel transfer problem. The TAF pilot could not re-start it and ejected at about 25 km southeast of Rhodes, being picked up half an hour later by a Greek helicopter, wet, cold and with some bruises on his face.

Despite their age HAF Mirage F1CG, air superiority fighters continue to keep a close watch on the Aegean Sea in company of the more modern Mirage 2000 and F-16 Fighting Falcons.

Libyan Mirage F1ED number 515 armed with two Magic AAMs intercepted over the Mediterranean by an US Navy F-14 Tomcat and photographed by its RIO. A great percentage of Libyan F1 pilots are believed to be French expatriates and Pakistanis. (US Navy)

The Libyan Arab Air Force is the only air arm, outside the SAAF, to have ordered the non-radar Mirage F1A, sixteen F1AD, also equipped with retractable flight refueling probes were acquired, these were the first to be delivered between January and April 1978. (Dassault)

The Egyptian AF, staring in 1974, has received 82 first-generation Mirages, including 54 Mirage 5SDE (illustrated), six 5SDDs, six 5SDRs and 16 5E2s, serving with two regiments at Tanta and Genaclis air base, mostly used in the ground attack role. These were used in the short border war with Lybia. Also Egypt operates a fighter regiment equipped with the survivors of 16 Mirage 2000EM and four Mirage 2000BM delivered in 1986-1988. (Cuny)

The Kuwaiti Air Force (KAF) acquired 33 Mirage F1s originally assigned to 18 and 61 Squadrons at Ali al Salem, the first batch of 18 F1CKs and two BK trainers were delivered in desert camouflage staring in July 1976, these were joined from December 1984 by nine F1CK2s and four F1BK2s with upgraded avionics and overall grey color scheme. 16 single seaters escaped to Saudi Arabia in August 1990, regrouping at Taif before flying 128 combat missions, with 200 flight hours -of a total of 634 and 700 respectively while in exile- from Dharan. No losses were sustained and between 1 and 3 July 1991 the Mirages returned to Ali al Salem, most still wearing the their "Free Kuwait" inscriptions. The F1s were retired in 1993, after the F-18C/Ds became operational. (Dassault)

(Above and Next) Morocco Mirage F1 order was placed in December 1975 in the form of 25 F1CH fighters and an option on 50 more. Roughly at the same time Morocco annexed the Western Sahara when this territory was vacated by Spain commencing a hard fought war with the Polisario forces, backed and armed by Algeria and Lybia. In 1977 half the option was exercised. Thus the Royal Air Force of Morocco RAFM received 30 Mirage F1CHs, 14 Mirage F1EHs and six F1EH-200 (the later with air-to-air refueling probes), with deliveries concluding in July 1982. These aircraft equip 5a BAFRA (Base Aérienne Force Royal Aérienne) at Sidi Slimane. Mirages were heavily involved in the war against Polisario guerrillas, until a provisional cease fire was signed, with at least three lost to groundfire, two in October 1981 and one in January 1985 all lost to surface to air missiles. Like Spanish Mirages, Moroccan F1s have received Tracor AN/ALE-40 flare-chaff dispensers. The illustrations show an F1CH, number 126, with an RP-35 1200-liter drop tank (34 - Dassault); two F1CHs flying in formation with two Spanish F1EEs from 462 Sqn. near Sidi Slimane during some joint exercises late in 1991 (via J.Bas) and an F1EH-200 number 175. (Dassault)

Hellenic Air Force Mirage F1CGs have received a modification which consists in wiring the underwing outer pylons, then fitting an Aero 3B launcher for Sidewinder infrared-guided AAMs, thus doubling its capability to four AIM-9P4 Sidewinders, including the two fitted in the standard wingtip launchers. (342 Sqn.)

Insignia worn by Greek Mirage F1 pilots. (Author)

Insignia of 334 "Talos" Mira Pantos Kerou.

Insignia of 342 "Sparta" Mira Pantos Kerou.

Pilots of 334 Squadron in company with their colleagues of 111 Squadron, Spanish AF at Manises-Valencia AB, during an exchange visit in June 1993. (Author)

334 Squadron re-deployed from Tanagra, some 60 km northwest of Athens, to Iraklion AB in the island of Crete, thus covering the southern approaches of the Aegean Sea. Greek Mirage F1s are painted in the original French air defence blue scheme. (Author)

41-year old Lt. Col. Alexandros "Al" Papagermanos, commanding officer of 342 Squadron, part of 114 Combat Wing, Hellenic Air Force. (Author)

Pilots of 342 Squadron with their CO, September 1995. (Author)

The Mirage F1CGs from 342 Squadron wear on the right front fuselage the names of Greek Islands, this aircraft carries the name of Samothraki (Samotrace). (Author)

With Captain Dimitrios Tsirogiannidis at the controls, this Mirage F1CG of 342 Squadron receives the last ground crew checks in the last chance area. (Author)

Mirage F1CG in landing configuration. Generous flaps make the F1 an easy aircraft to land, although its narrow landing gear could be a little troublesome in crosswind landings. (Author)

Despite being between seventeen and twenty years old, Hellenic Mirage F1s are maintained in very good shape thanks to the arduous efforts of the maintenance people and Hellenic Aerospace Ltd where general overhauls are performed. (Author)

Mirage F1CG "Naxos" with its pilot aboard. (Author)

Two 342 Sqn. Mirage F1CGs prior to an air combat tactics mission, note the centerline RP-35 1200-liter drop tanks. (Author)

Lt Nicolaos Sialmos of 342 Squadron, lost with Mirage F1CG serial 116 "Lasvos" in the Aegean Sea after engaging a pair of intruding Turkish F-16s on June 18, 1992. (342 Sqn.)

IN THE SHADOW OF DESERT STORM

For the first time since its creation, NATO ordered forces under its direct command to take part in a major conflict. Operation "Attic Front" was launched to deter Iraq from attacking Turkey, a member of the Alliance.

Belgium originally procured 106 Mirages comprising 63 single-seat Mirage 5BAs, 16 Mirage 5BD trainers and 27 Mirage 5BR recce aircraft. The first example of each variant was built by Dassault, but the remainder were license built by SABCA at Gosselies, Belgium. At its peak the type equipped four squadrons and for many years was the backbone of the Force Aérienne Belge tactical ground attack and recce force, replacing the Republic F-84F Thunderstreak and RF-84F Thunderflash, with the first Mirage 5BA entering service in July 1970 with 8 Escadrille (also known in Flemish as 8 Smaldel. Other units with ground attack Mirages were 1 and 2 Escadrilles (both reequipped with F-16 Fighting Falcons in 1987). In January 1991 ten aircraft from 8 Escadrille and five from 42 Escadrille deployed to Diyarbakir air base (also known as Erhac) in Turkey to bolster Coalition forces in case of an Iraqi attack on that country during Desert Storm (see insert). The aircraft returned to Florennes after the war and 8 Escadrille disbanded on 13 September 1991, leaving only 42 Escadrille with the Mirage, with an inventory of eleven Mirage 5BAs, 18 Mirage 5BRs and five Mirage 5BDs. The intention was to maintain this squadron, based at Bierset, with a mix of upgraded MirSIP (Mirage Systems Improvement Programme) and recce aircraft, but budgetary cuts implemented by the Belgian government forced its disbandment, which took place on 22 December 1993. (de Bruyn)

ACE Mobile Force

AMF (Allied Command Europe Mobile Force) was set up in 1960 with the specific task of rapidly reinforcing vulnerable areas of the Alliance following NATO's "flexible response" doctrine, with the anticipated attacker being the very potent Warsaw Pact forces. Two of these mobile conventional deterrence forces have been created, dedicated respectively to the northern and southern parts of Europe. They are formed by air and ground units from Belgium, Canada, Germany, Italy, Luxembourg, the Netherlands, the U.K., and the U.S., and regularly hold large maneuvers to measure the effectiveness of the AMF.

Five out of seven AMF contingency areas are situated in Southern Europe, three of which are in Turkey. The air component of AMF South consists of aircraft from Belgium's No.8 Sqn. with Mirage 5, German Alpha Jets of the Oldenburg based JaBoG 43 and RF-104G of the Italian 132 Grupo from Villafranca, while the land component has helicopter support from Germany (UH-ID) and the U.K. (Puma).

A Mirage 5BR lands at Bierset in the classic attitude of delta Mirages, note that Belgian BAs and BRs were retrofitted with a Loral Rapport III ECM system, identifiable by the bulge at the base of the vertical fin. (Soupart)

(Above and Below right) Mirage 5BA (serial BA54) was one of the veterans of the Diyarbakir deployment and was aptly named "Diyarbakir Express" painted in white over the left side of its nose. (Soupart)

Operation Attic Front

On the eve of 1990, only a month away from the dead-line of Iraq's unconditional retreat from Kuwait imposed by UN resolutions, a war in the Middle-East seemed to be inevitable. Anticipating the risk of getting militarily involved by complying with the UN economic embargo and as such shutting down Iraq's important oil pipeline which runs over its territory, Turkey decided to call in the air asset of NATO's AMF Southern component. The aim was obviously to deter Iraq, sharing a border of some 250 Km with Turkey, from opening a second front in the north, and to help defend sovereignty in case of eventual hostilities

The call for assistance from the Turkish government was well received by NATO on 20 December, and after consultation of the nations involved, operation Attic Front was born and the deployment of 18 Alpha Jet and 6 RF-104 to Erhac and 18 Mirages to Diyarbakir was ordered. Their mission consisted of conducting strictly defensive operations in Eastern Turkey.

Although the sending of 42 fighters seemed little more than a gesture of good will, this decision had important consequences for all parties. Clearly, Iraq was warned not to challenge NATO, which was willing and ready to take whatever additional steps it found necessary, and, in contrast to the Desert Shield alliance, Iraq's propaganda aimed at dividing the coalition would surely have no impact on NATO. Furthermore, it kept important numbers of Iraqi Mirage F-1 and Mig-29 Fulcrum fighters as well as army divisions, including Republican Guards, some 1500 Km from the KTO (Kuwaiti Theater of Operations).

A Mirage 5BR from 42 Escadrille at its hardened aircraft shelter at Bierset air base, Belgium, July 1993, the squadron disbanded five months later, on December 22, and the Belgian AF (BAF, or Force Aérienne Belge FAB) was left since then with its F-16 Fighting Falcons as its only combat aircraft. (Soupart)

Belgium's Participation

A few weeks before the official announcement to send Mirages to Turkey, a short political quarrel regarding the eventual deploy-

A section of 42 Escadrille Mirage 5BAs flying in close formation over typical Belgian landscape, noticeable are the 1700-liter drop tanks and the wrap-around camouflage. (de Bruyn)

BA54 "Diyarbakir Express" being refueled from a fuel bowser at Bierset. (Soupart)

Mirage 5BR (BR22) from 42 Escadrille caught during a CAP mission over the Anatolie area, southeast Turkey, late in January 1991, the aircraft is armed with two live AIM-9N Sidewinders and full ammunition for its two DEFA 552 30mm cannon. (de Bruyn)

One of the two Mirage 5BD two-seaters which served with 42 Escadrille until the unit disbanded in December 1993. (Soupart)

ment of Belgian F-16s to assist in operation Desert Shield highlighted the vulnerability of these aircraft as they are not yet equipped with RWR or ECM systems. However, besides having an active/passive ECM fit, the fact that the government decided to send the Mirage 5 was no surprise to the air force community, as the Bierset-based 8 Sqn. is the BAF's dedicated AMF unit.

Only a day after the government's "go-ahead" on January 2nd, the movement of support personnel and equipment to Diyarbakir had already started in order to commence immediate operations upon arrival of the Mirages, shelled for the 7th. Up to four flights a day were made by BAF C-130s of the 20 sqn. out of Bierset and Melsbroek, moving large amounts of equipment, ranging from winter clothing to field kitchens and live ammunition to trucks and aircraft tractors. Everything necessary to operate a squadron of combat aircraft some 3200 Km from home was transported by air, in contrast to previous AMF exercises where most of the heavy equipment was shipped to Turkey in order to reduce costs.

Counting numerous "veterans" of these earlier deployments amongst the ground staff of the 3rd Wing, the setting-up of ground support was handled without any major problems, especially as Diyarbakir was not an unknown base, with two AMF exercises having already been conducted from this base in '83 and '87 .

Before their departure, combat pilots spent many hours attending non-routine briefings such as Rules of Engagement, NBC, desert and winter survival, etc., besides being updated on the military situation in the area of operations, although the squadron was not unprepared as their flying maps of the Warsaw Pact area had already been replaced by one of Iraq months before! Pilots had not forgotten that "detente" did not mean that they did not have a job to do.

The first and second stage of the Mirage ferry-flight, starting on Sunday, January 6th, were conducted without major problems. The four-ship formations took off from Bierset in half-hour intervals, heading for Amendola in Italy to refuel, before continuing the next day to Balikesir in Turkey. However, weather conditions in Turkey proved to be harsh, and this scheduled stop turned out to last almost a week, imposed by thick fog and snowfall at both Balikesir and the final destination of Diyarbakir. The BAF Boeing 727 of the 21 Sqn departing Bierset on Monday 7th with some 100+ personnel on board joined the combat aircraft at Balikesir, but was also trapped for two days, as were several C-130s. As such, Balikesir became a transloading base for tons of cargo, inflicting a review of the keenly developed transport plan, a situation which eventually came to an end five days later.

At Diyarbakir, so far no real threat was present as war had not started yet, and initial routine missions consisting of area familiarization flights and navigation sorties, some in the company of aircraft from the local 182 Filo (Sqn.) equipped with the CF-104 or the 184 Filo with their RF-5As, were on the order of the day. However, this was not a routine AMF. Besides the presence of Mürted-based Turkish F-16Cs which had replaced the 181 Filo F-104Gs, on their turn deployed to nearby Batman airbase, the SHORAD (Short Range Air Defense) base defense with AAA and Rapier missile batteries was activated. As the deadline date of January 15th approached, all Mirages were also ready to commence defensive actions if required. For the first time in their ca-

**Ironically, the deployment order came shortly after the official announcement that the 8 Sqn would disband before the end of 1991!*

The nose contours of the Mirage 5BR can be seen in this close-up picture of BR15. (Soupart)

Despite the disbanding of 42 Escadrille, SABCA continued with the MirSIP program, upgrading fifteen Mirage 5BAs and five Mirage 5BD, these eventually being sold to Chile for reequipping Grupo 8, replacing Hawker Huters and becoming Mirage 5MA, and 5MD "Alkan", the contract also including four unmodernized Mirage 5BRs and one Mirage 5BD. The Mirsip Mirages have a new nav/attack system integrated by Sagem of France. (Soupart)

42 Escadrille insignia.

reer, they carried two live AIM-9N Sidewinder AAMs in addition to the 30mm war ammunition for the internal guns.

From Attic Front to Ace Guard: The start of the war

With the coalition air offensive against Kuwait and Iraq starting in the early morning of January 17th, tension rose! Getting involved in the war became even more imminent a day later after the first Iraqi Scuds were targeted against Israel. Indeed, Diyarbakir was well in range of this missile type, as was Incirlik, the base used by the USAF to conduct offensive raids against Northern Iraq. Snakeye, Rockeye, and Durandal equipped CF-104s were kept on alert and F-16s rotated day and night manning border CAPs Combat Air Patrol), while the Belgian Mirages, performing daytime navigations, were ready to intervene whenever called upon by NAEWF AWACS aircraft.

With the continuing success of the coalition air raids resulting in little to no serious reply from the Iraqi side against them, the threat against Turkey was assessed to be decreasing, hence more routine type missions were flown, although weapons configuration remained unchanged. Good use was made of the areas of low population to train low-level valley-flying in the mountainous region to the north of Diyarbakir and daily practice bomb runs were made using the range located within the base perimeter as well as the Borankoy range close to Erhac.

Traffic at Diyarbakir remained intense throughout the whole period with a large variety of aircraft types transiting. Besides the resident Base Flight T-33s, C-47s and SAR UH-IHs, an Army UH-IH and calibration DO-28Ds used Diyarbakir as temporary base, while Turkish F-4 Phantoms from Erhac and Eskisehir visited frequently. Overflying Diyarbakir to and from the war zone, US warplanes having technical problems or being damaged in combat diverted from time to time before returning to Incirlik after being repaired. Obviously, transport aircraft made up the largest part of the visiting aircraft, with Turkish C-130 and C-160, Dutch F-27, German C-160, US C-130 and C-12, Belgian C-130 and Boeing 727 as well as a variety of Russian built transports, the latter reflecting the changing mood in the world. Another prove of detente was the fact that Belgian Hercules' and B.727s were occasionally granted overflight of Czechoslovakia, Hungary, Romania and Bulgaria, effectively reducing their flight time while increasing available payload.

Base Defense

Although close-in base defense was provided by the already mentioned Turkish Rapier short range missiles and anti-aircraft guns, additional equipment was flown in. Scoring high rates of success in destroying incoming Iraqi Scud missiles in Saudi-Arabia and Israel, two Patriot squadrons of the Dutch Air Force were highly welcomed at Diyarbakir. However, one time they scared the hell out of everybody when two missiles were actually launched following the acquisition of a wrong target, auto-destroying a few seconds later.

Eager to contribute their part in the Gulf crisis, even more Dutch air defense was brought in. Next to follow was a battery of Improved Hawk and teams equipped with Stinger portable anti-aircraft missiles. Since the Dutch Air Force has no heavy-lift transport aircraft, all personnel and some cargo was transported to Diyarbakir by RAF, BAF, and GAF transports, while the heavy equipment was moved by chartered civil Il-76 Candid and An-124 Condor aircraft. These USSR registered airlifters started to appear regularly and added an 'exotic' touch to the already important traffic, although even more Eastern-Block transports would arrive on the scene.

Having experienced much public opposition when attack aircraft were deployed to Turkey, Germany decided, partly to comply with U.S. requests to play a more active part in the crisis, to contribute to the air defense of Turkey. About a month after hostilities commenced in the Gulf, highly mobile Roland, as well as additional Hawk missiles arrived at Diyarbakir. The heavy support equipment for the Hawk and the large trucks on which the Roland systems are mounted arrived in USAF C-5s, although an AN-124 was chartered to do the job. However, the pilot refused to fly the already loaded aircraft to a war zone, making the use of MAC's super-transports indispensable but delaying the deployment for several days. Lighter cargo was transported by Luftwaffe C-160s, while the personnel was flown in with B.707 and ex-GDR TU-134s and TU-154s.

Shoulder insignia carried by Belgian Mirage 5 pilots.

With the start of the allied ground offensive and its rapid success the end of the war was nearing faster than anticipated. Although it was assumed that the presence of the AMF was required even after the liberation of Kuwait, the announcement by NATO that the squadrons were to return home the following week was received with great relief. Ir, all, the Mirages flew 960 hours during operation Attic Front/Ace Guard, the last fighter arriving back at Bierset on March 12th. The 15 Wing for their part accumulated some 870 Hrs during 61 flights of logistic support for the deployment.

Although the aircraft saw no combat action, the mere presence of the AMF and other units showed that NATO was willing to do what it was meant to do in a crisis situation; the collective defense of the Alliance.

(Above and next page) In 1976 nineteen Mirages were delivered to the kuwaiti Air force, followed in 1983 with eleven more, thus a total of 27 F1CK/CK2 and seven BK/BK2 were in service with KAF, coded 701 to 727 and 771 to 776 respectively. Actually, fifteen Mirages remain stored in new air conditioned hangars at the Ali Al Salem air base and are up for sale. Eight Mirages were lost in accidents since 1976. By early 1996 five F1BKs and three F1CKs remained in Iraq after they were looted during the Emirate invasion in August 1990. Mirages served with 18 and 61 Squadrons at Ali Al Salem, and during the Iraqi invasion shot down two Su-22 and two Mi-24, all with Matra R550 Magics. Most of the Mirages escaped to Taif, Saudi Arabia, and later moved to Daharan. From there and during "Desert Shield", the fouteen F1s which were available made 429 missions and 425 flight hours, while in "Desert Storm" recorded 205 missions and 275 flight hours. After the war the Mirages returned to Kuwait, being based at Kuwait City International airport, until they were phased out of service in 1993 and the 16 and 81 Squadrons disbanded, when the F/A-18s were declared operational. But, although stored in very good condition, the Mirages are flown once or twice a week by two KAF senior officers to keep up their flying time, one of them being Bigadier General Saber M. Al-Suwaidan, the present commaning officer of the KAF. (KAF via S. Mafé)

Mirage F1CK 715 crash landed after landing gear failure at the International base in 1992. The aircraft is a write-off and is being rebuilt for the KAF Museum. (KAF via S. Mafé)

Armée de l'Air Mirage 2000C fighters participate in Deny Flight operations since its inception, with a six aircraft detachment based at Cervia, Italia, illustrated is a Mirage 2000C (RDI) from Escadron de Chasse 1/12 "Cambresis" returning from a CAP sortie. (Sirpa Air)

12

Close-Ups

Swiss Mirages are wired for JATO bottles when operating from short mountain airfields, illustrated in spectacular fashion by this Mirage IIIRS, one of eighteen received by the Swiss AF and operated by 18 Staffel (Squadron), one of the permanent units assigned to the Surveillance Wing. (F+W)

The Swiss Mirage IIIS are among the more original of any variants, having replaced the Cyrano IIbis fire control system, by the American developed Hughes Taran 1, while the MATRA 530 AAMs were supplanted by two AIM-4 Falcon missiles, as well as the two AIM-9 Sidewinders. (Swiss AF)

In French Armée de l'Air service, the Mirage IIIEs (the last were phased out in 1994) were true multirole fighters, and this picture clearly demonstrates by the armament carried. From left to right, Martel anti-radar missile; Matra R530 AAM, and AN52 tactical nuclear bomb. (Dassault)

An RAAF Mirage IIIO assigned to the Air Research and Development Center (ARDU), photographed in the mid-eighties, while performing trials with Paveway II laser guided bombs, in this case two GBU-16s attached to the combined fuel tank and weapons carrier. (Texas Instruments)

The Dassault Mirage 5 was developed upon an Israeli Air Force request for a relatively cheap fair weather fighter bomber, without the —then— sophisticated and troublesome Cyrano fire control radar, at the end the order for 50 Mirage 5J was embargoed, and these fighters went into Armée de l'Air service as Mirage 5Fs, while the Israelis made an almost clonic copy, the Nesher, which later was developed into the rather more powerful Kfir. Here the Mirage 5 prototype shows its weapons capabilities, in the foreground is an AS 30 air-to-surface missile. (Dassault)

For maintenance the Mirage IIIE's Cyrano IIbis is removed together with the radome forming a compact and easy transportable subject. (author)

Two Mk.106 practice bombs fitted on an Alkan 65 bomb carrier in an Argentine Mirage IIIEA. (Padin)

The Matra Phimat chaff dispenser pod fitted on the outer wing pylon in an Armée de l'Air Mirage IIIE. (author)

A Thomson-CSF Barracuda II jamming pod carried by a French Mirage IIIE. (Author)

The camera fit in a recce Mirage IIIR of 18 Staffel, Swiss AF, note that this is an upgraded Mirage, noticeable by the canard foreplanes and RWR antennas on wingtips and fin. (Boisselon)

Instrument panel of a Mirage IIIEE. (Author)

Instrument panel, front seat of a twin-stick Mirage IIIDE, note the lack of radar display screen, as in the single-seat. (Author)

Close up of the cockpit area of a Belgian Mirage 5BD, showing clearly the zero-zero Martin-Baker Mk.10 ejection seats, and the typical supersonic conical intake, which was followed in the Mirage F1 and 2000. (Soupart) 12/

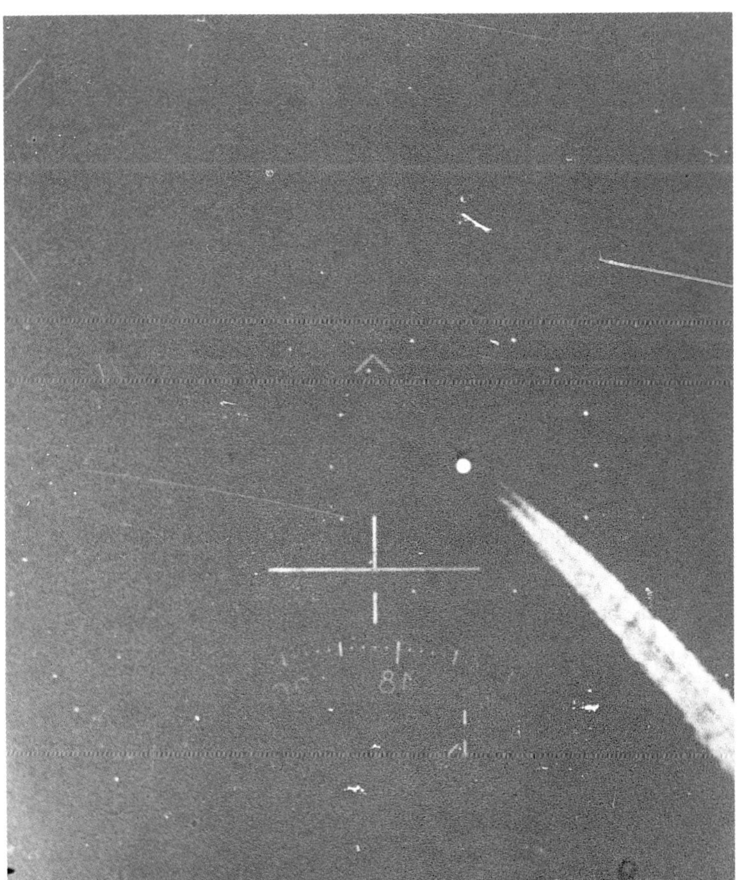

FOX 2! - During an DACM hop, a Mirage IIIEE of Spanish AF's 111 Sqn., gets into firing parameters on an F1CE from 141 Sqn., as seen through the Thomson-CSF 97 HUD, the pipper is on the F1, while distance is slightly over 1800 meters, in good firing parameters for the AIM-9 Sidewinder. (111 Sqn.)

Armorers loading a MATRA R550 Magic 1 into a Grupo 6 de Caza, Argentine Air Force, Mirage IIIEA. (Padín)

Instrument panel of the Kfir C1, on the top right is the radar warning receiver display. (IAI)

Instrument panel of the Kfir C7, a noteworthy difference is the digital display indicator on the lower left of the panel. (IAI)

Instrument panel of the Belgian Mirage 5 upgraded under the MIRSIP program, of which fifteen aircraft have been sold to the Chilean Air Force as the Alkan. (Sagem)

Kfir C1 weapons capabilities, including SUU-23 20mm cannon pods. (IAI)

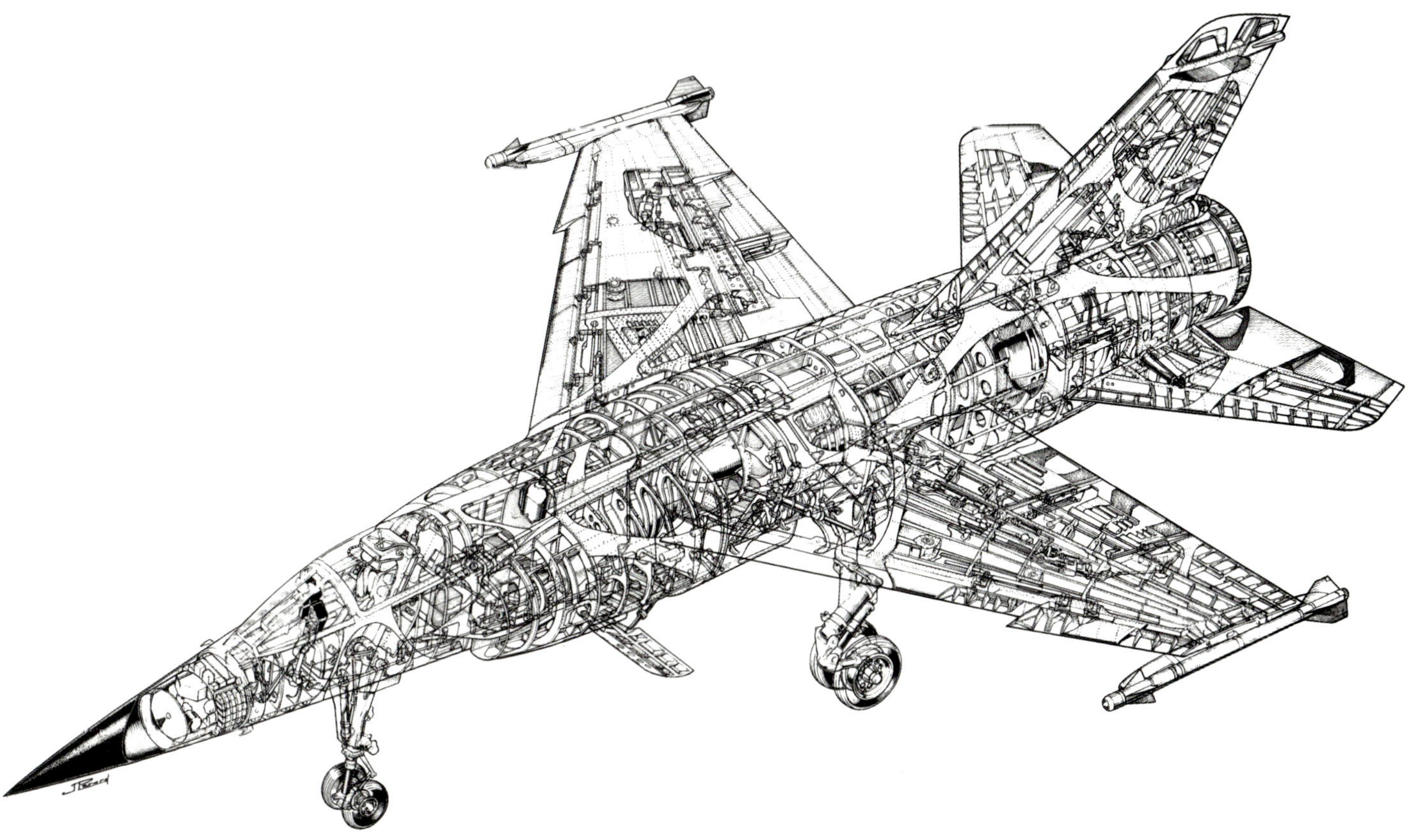

Cutaway drawing of a Mirage F1. (Dassault)

Close up of a Spanish Mirage F1CE, in foreground is a captive AIM-9N3 Sidewinder AAM. Note below the horizontal stabilator the Tracor AN/ALE-40 flare-chaff dispenser, one on each side. (author)

The AN/ALE-40 regulating box is located behind the ejection seat. (author)

The control set for the AN/ALE-40 inside the cockpit of a Spanish F1. (author)

The Atar 9K-50 turbojet which powers the Mirage F1 series, as well as some Mirage III, 5 and the Cheetah. (Author)

Martin-Baker Mk.6 (zero-zero) ejection seats which equip Spanish Mirage F1CEs and EEs. (Author)

Spanish Mirage F1s are being equipped with an advanced AN/ALR-300 radar warning receiver, notice the fin antennas, in place of the older Thomson-CSF BF. (Author)

A Spanish AF armorer loading 500-lb Mk.82s into a Mirage F1EEDA from 111 Squadron, the centerline pylon can load four bombs. (Author)

Instrument cockpit of a Spanish Mirage F1EE retrofitted with the AN/ALR-300, its display is located on the top left of the instrument panel. (author)

The Mirage F1CT is the latest French variant of this classic fighter made by retrofitting earlier F1Cs, optimized for ground attack missions, this examples just rolled out from the conversion plant has not been camouflaged yet, note the 2200-lit. centerline tank. (Soupart)

Mirage F1EDA from 111 Squadron loaded with six Mk.82. (Author)

This Mirage F1CE from Ala 14 carries two Matra R530Es radar guided AAMs, two Sidewinders and a centerline 1200-liter drop tank. (Calvo)

Close up of a Mirage F1EDA from 111 Squadron in which can be seen a Matra Super 530F-1 radar guided air-to-air missile, this AAM can reach targets flying at almost 80.000 ft and has an amazing capability for snap-up and snap-down shots. (Author)

A Paveway II (GBU-10) 2000-lb laser guided bomb being driven for loading into a Mirage F1, note that the nose guidance unit has not been fitted yet. (Author)

Close-up of an Armée de l'Air Mirage 2000B conversion trainer showing the cockpit area, the Mk.10 seats and the small foreplanes over the intakes. (Soupart)

A Spanish Mirage F1 taking orf in full afterburner carrying a light load, two Mk.20 Rockeye II cluster bombs. (Author)

Close-up of the Mirage 2000C fighter, note the fixed AAR probe, the mock cockpit, and the tiger stripped pilot's helmet. (Soupart)

Fin area of a Escadron de Chasse 1/12 "Cambresis" Mirage 2000C, note the Thomson-CSF Several(?) radar warning receivers fitted on the leading and trailing edges of the fin. (Soupart)

The Mirage 2000-5 is the latest variant produced by Dassault, with a new radar, Thomson-CSF RDY, in place of the earlier RDM and RDI, and new medium range "fire and forget" radar guided missiles, the Matra Mica. (Dassault)

Appendix A: Mirage IIIE Flight Manual

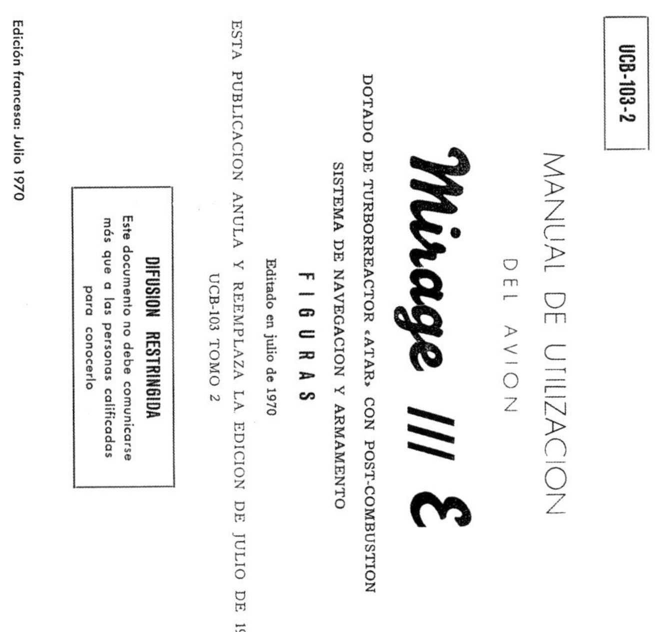

UCB-103-2

MANUAL DE UTILIZACION

DEL AVION

Mirage III E

DOTADO DE TURBORREACTOR «ATAR» CON POST-COMBUSTION

SISTEMA DE NAVEGACION Y ARMAMENTO

F I G U R A S

Editado en julio de 1970

ESTA PUBLICACION ANULA Y REEMPLAZA LA EDICION DE JULIO DE 1965-UCB-103 TOMO 2

DIFUSION RESTRINGIDA

Este documento no debe comunicarse más que a las personas calificadas para conocerlo

Edición francesa: Julio 1970

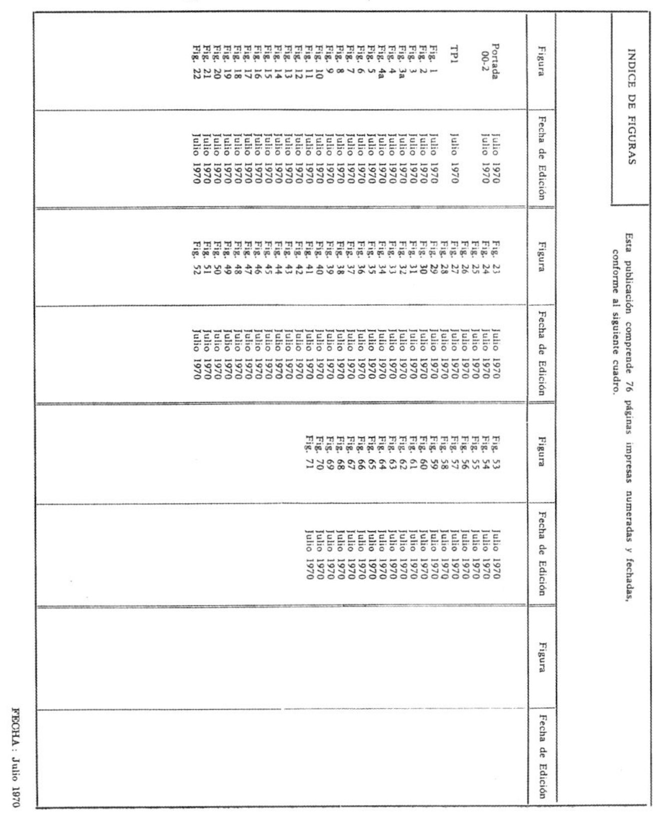

00-2 MIRAGE III E S N A

INDICE DE FIGURAS

Esta publicación comprende 76 páginas impresas numeradas y fechadas, conforme al siguiente cuadro.

Figura	Fecha de Edición	Figura	Fecha de Edición	Figura	Fecha de Edición	Figura	Fecha de Edición
Portada	Julio 1970	Fig. 23	Julio 1970	Fig. 53	Julio 1970		
00-2	Julio 1970	Fig. 24	Julio 1970	Fig. 54	Julio 1970		
		Fig. 25	Julio 1970	Fig. 55	Julio 1970		
		Fig. 26	Julio 1970	Fig. 56	Julio 1970		
TP1	Julio 1970	Fig. 27	Julio 1970	Fig. 57	Julio 1970		
		Fig. 28	Julio 1970	Fig. 58	Julio 1970		
Fig. 1	Julio 1970	Fig. 29	Julio 1970	Fig. 59	Julio 1970		
Fig. 2	Julio 1970	Fig. 30	Julio 1970	Fig. 60	Julio 1970		
Fig. 3	Julio 1970	Fig. 31	Julio 1970	Fig. 61	Julio 1970		
Fig. 3a	Julio 1970	Fig. 32	Julio 1970	Fig. 62	Julio 1970		
Fig. 4	Julio 1970	Fig. 33	Julio 1970	Fig. 63	Julio 1970		
Fig. 4a	Julio 1970	Fig. 34	Julio 1970	Fig. 64	Julio 1970		
Fig. 5	Julio 1970	Fig. 35	Julio 1970	Fig. 65	Julio 1970		
Fig. 6	Julio 1970	Fig. 36	Julio 1970	Fig. 66	Julio 1970		
Fig. 7	Julio 1970	Fig. 37	Julio 1970	Fig. 67	Julio 1970		
Fig. 8	Julio 1970	Fig. 38	Julio 1970	Fig. 68	Julio 1970		
Fig. 9	Julio 1970	Fig. 39	Julio 1970	Fig. 69	Julio 1970		
Fig. 10	Julio 1970	Fig. 40	Julio 1970	Fig. 70	Julio 1970		
Fig. 11	Julio 1970	Fig. 41	Julio 1970	Fig. 71	Julio 1970		
Fig. 12	Julio 1970	Fig. 42	Julio 1970				
Fig. 13	Julio 1970	Fig. 43	Julio 1970				
Fig. 14	Julio 1970	Fig. 44	Julio 1970				
Fig. 15	Julio 1970	Fig. 45	Julio 1970				
Fig. 16	Julio 1970	Fig. 46	Julio 1970				
Fig. 17	Julio 1970	Fig. 47	Julio 1970				
Fig. 18	Julio 1970	Fig. 48	Julio 1970				
Fig. 19	Julio 1970	Fig. 49	Julio 1970				
Fig. 20	Julio 1970	Fig. 50	Julio 1970				
Fig. 21	Julio 1970	Fig. 51	Julio 1970				
Fig. 22	Julio 1970	Fig. 52	Julio 1970				

FECHA: Julio 1970

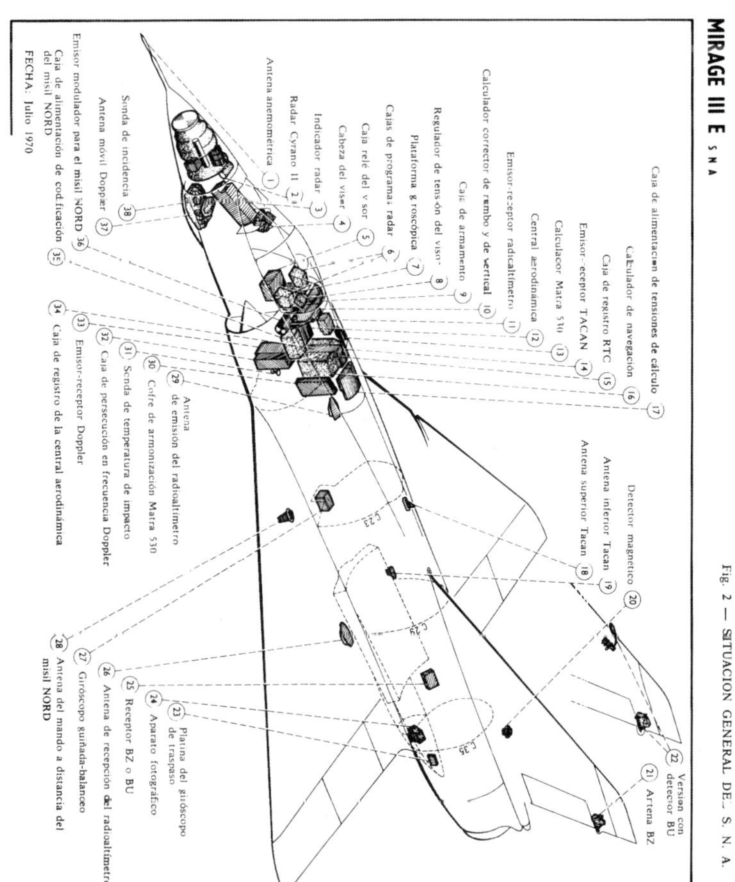

General arrangement drawing.

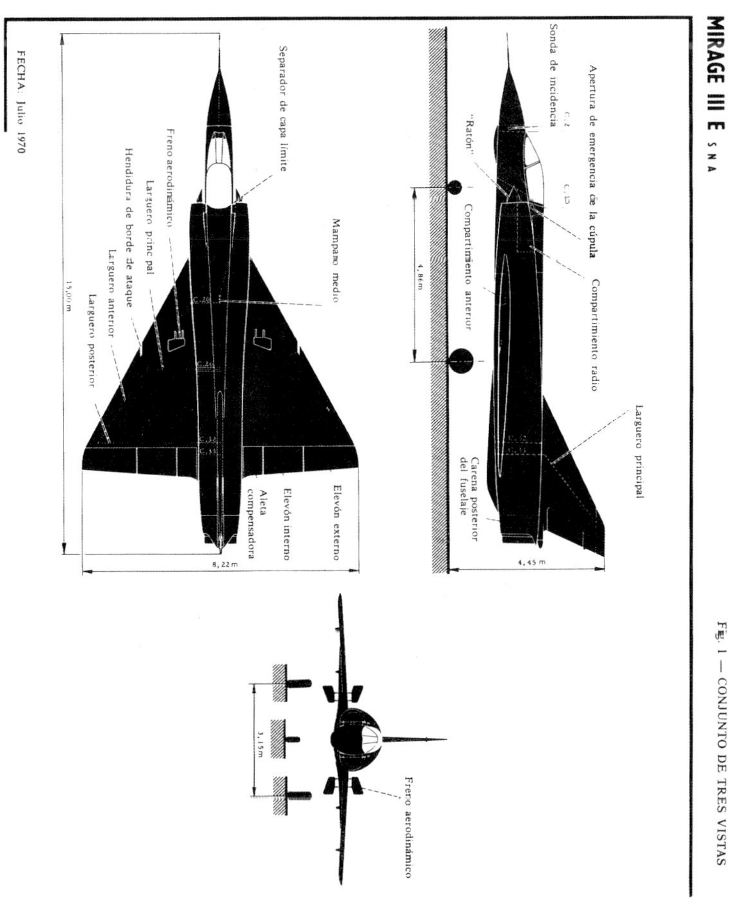

Three view arrangement drawing.

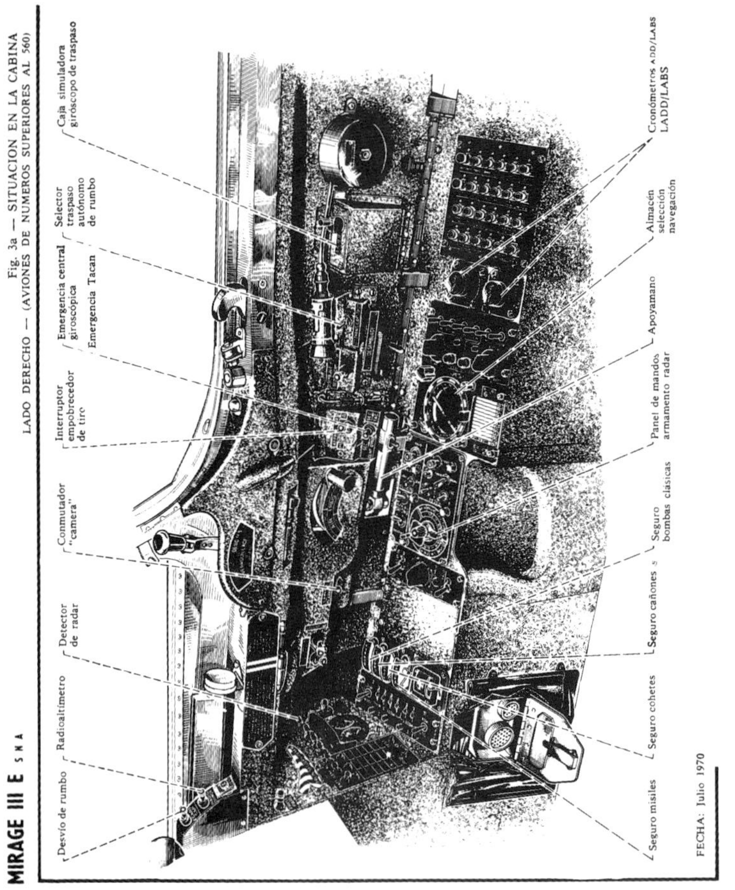

Right side controls.

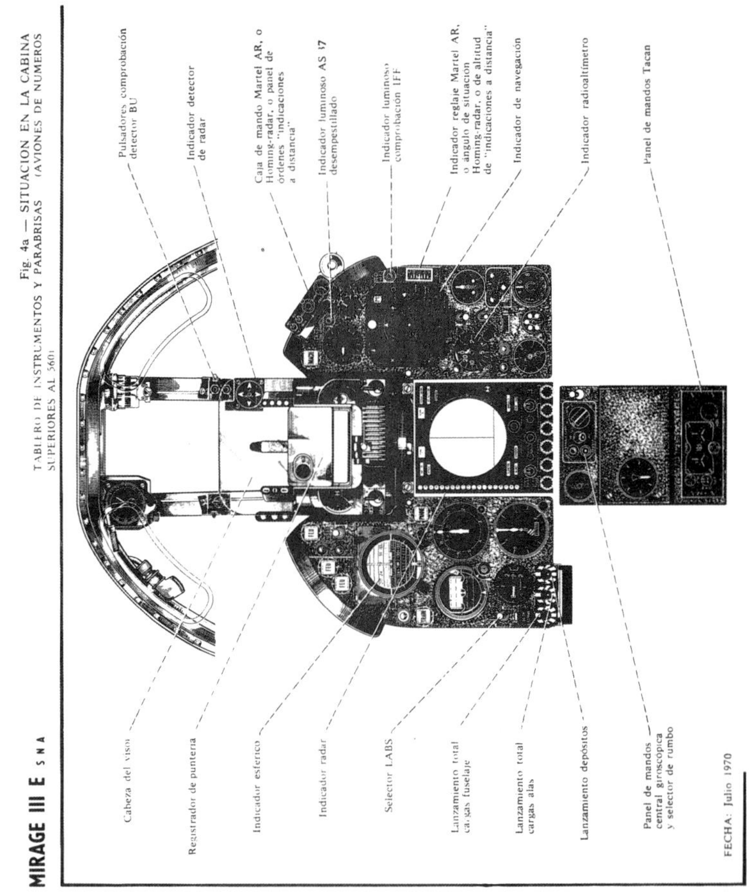

Instrument panel.

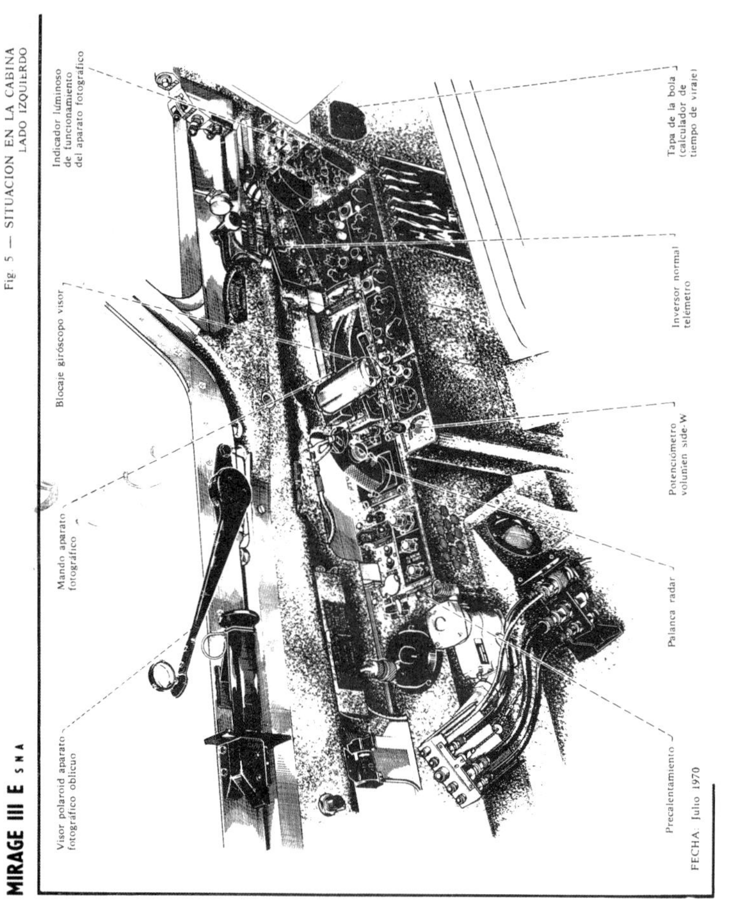

Left side console.

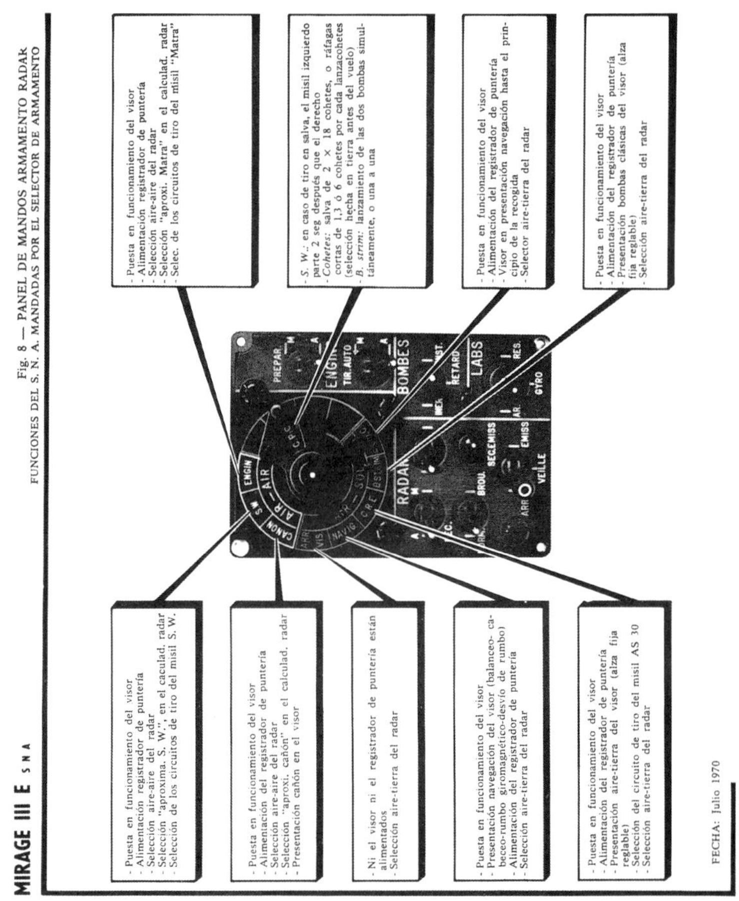

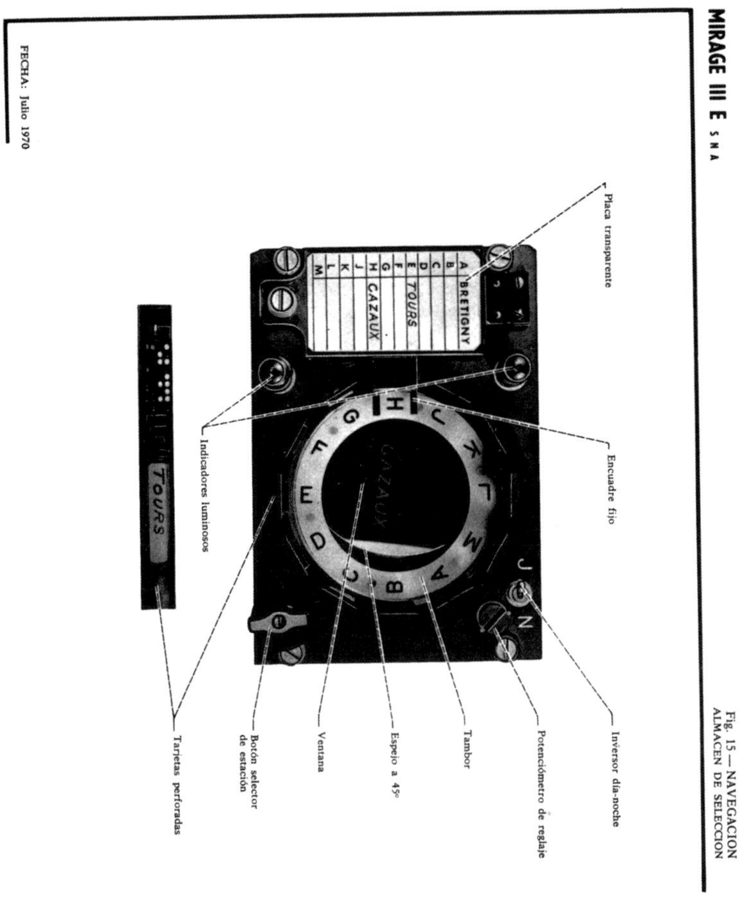

Doppler navigation system selection box with destination cards.

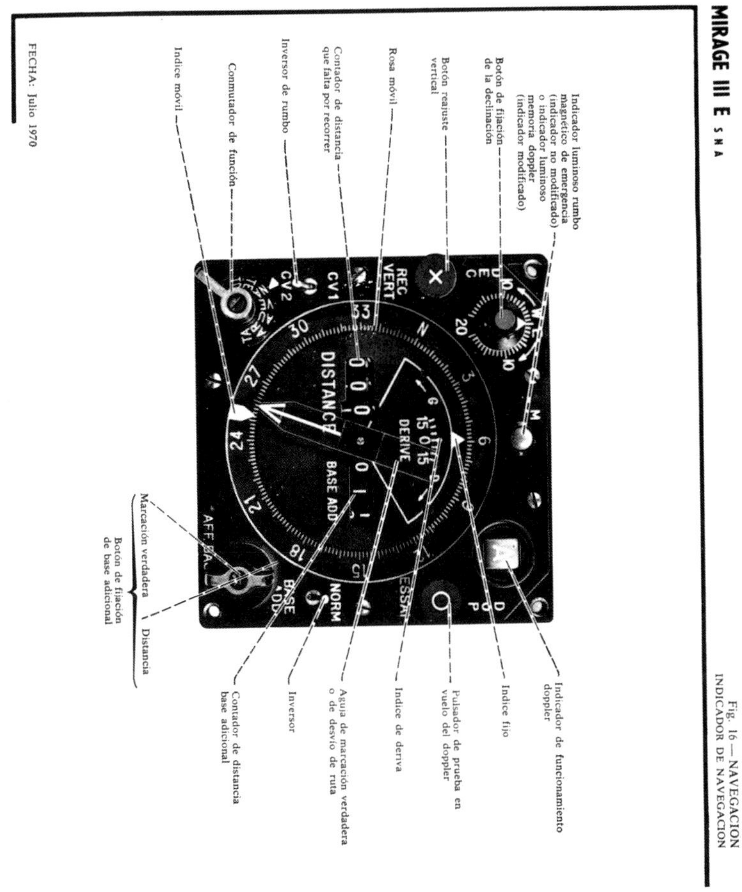

Doppler navigation system indicator on instrument panel.

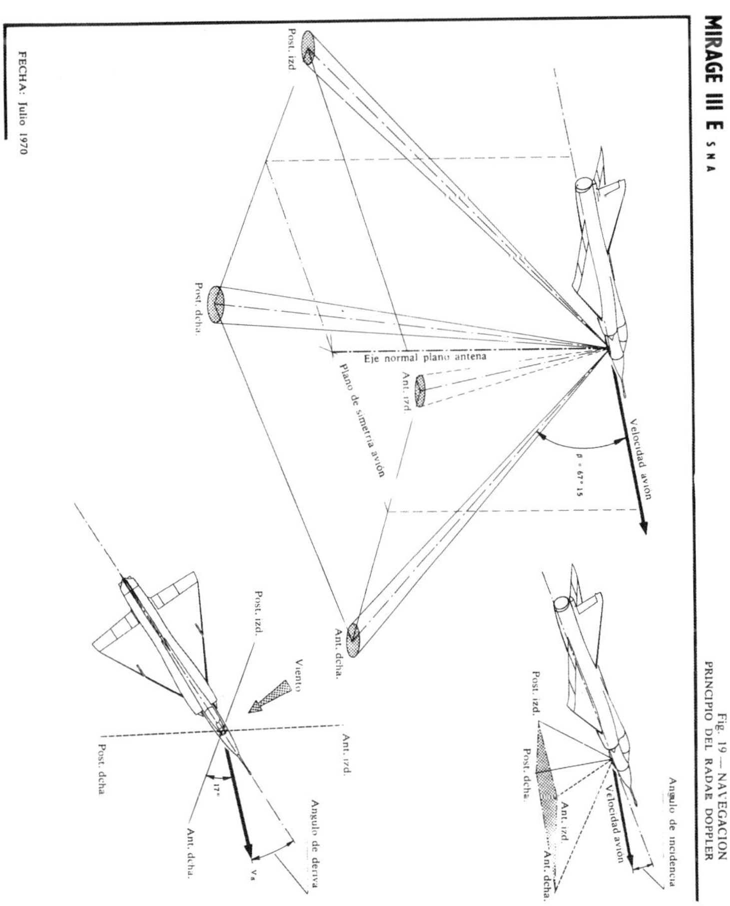

Working principle doppler radar.

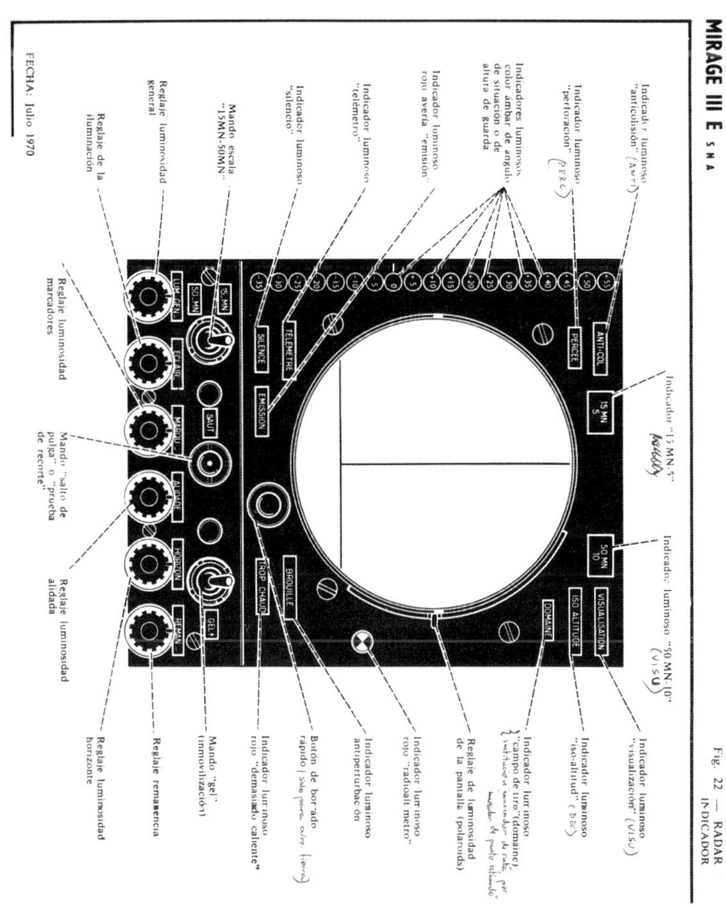

Cyrano IIbis radar display.

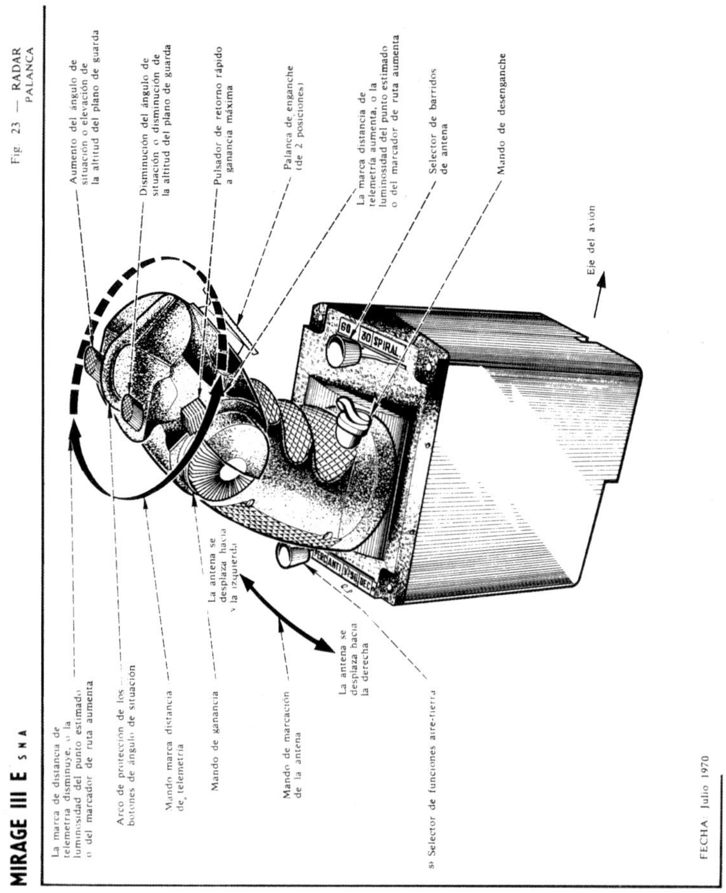

Radar joystick.

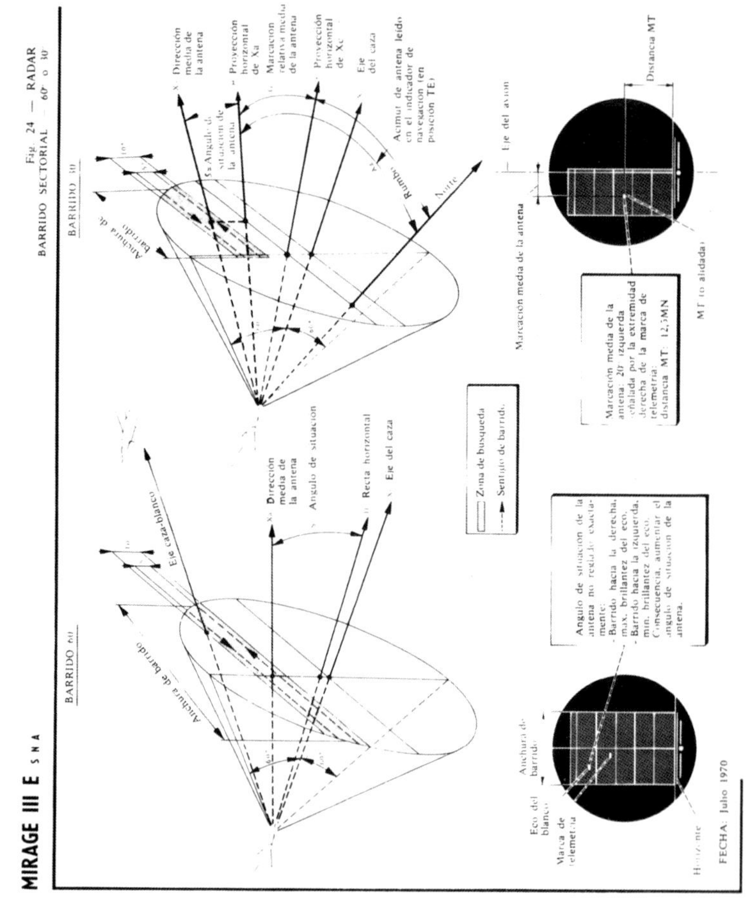

Cyrano IIbis sectorail sweep mode.

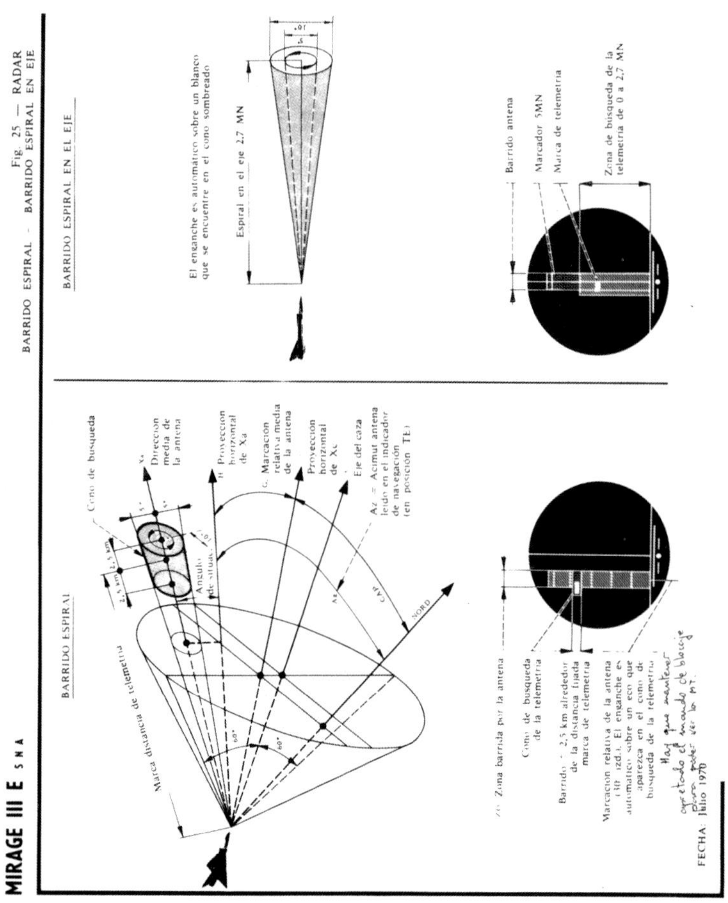

Cyrano IIbis radar spiral sweep and spiral axis sweep.

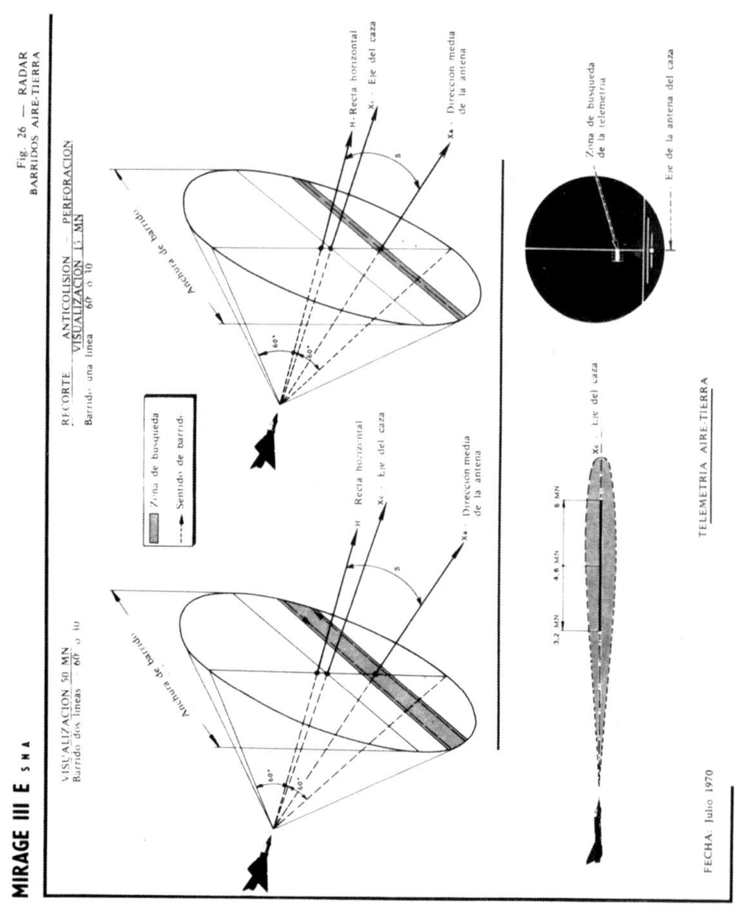

Cyrano IIbis radar air-to-air sweep.

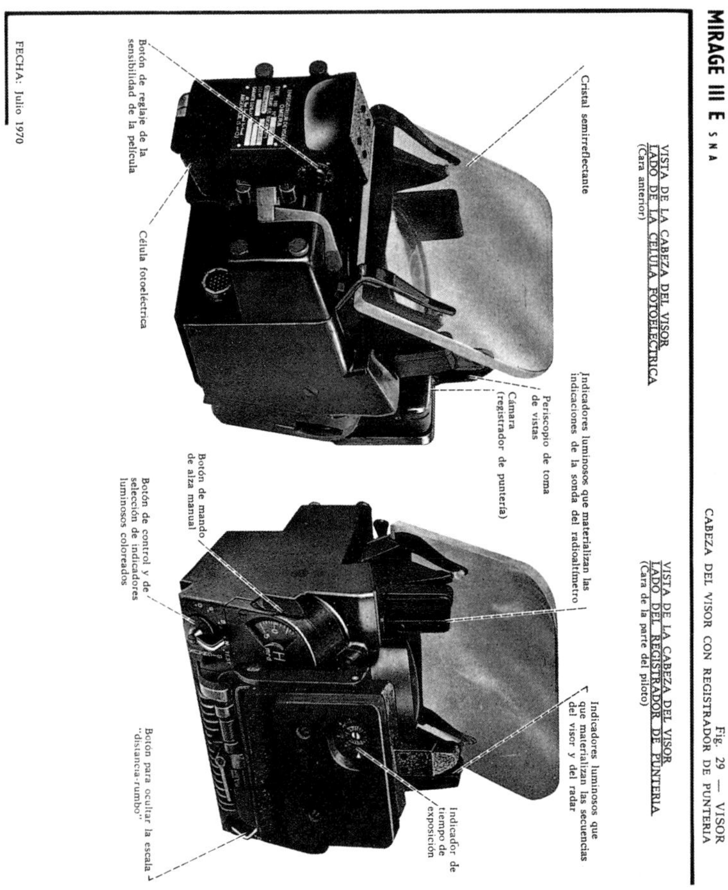

Thomson-CSF 97 description and working principle.

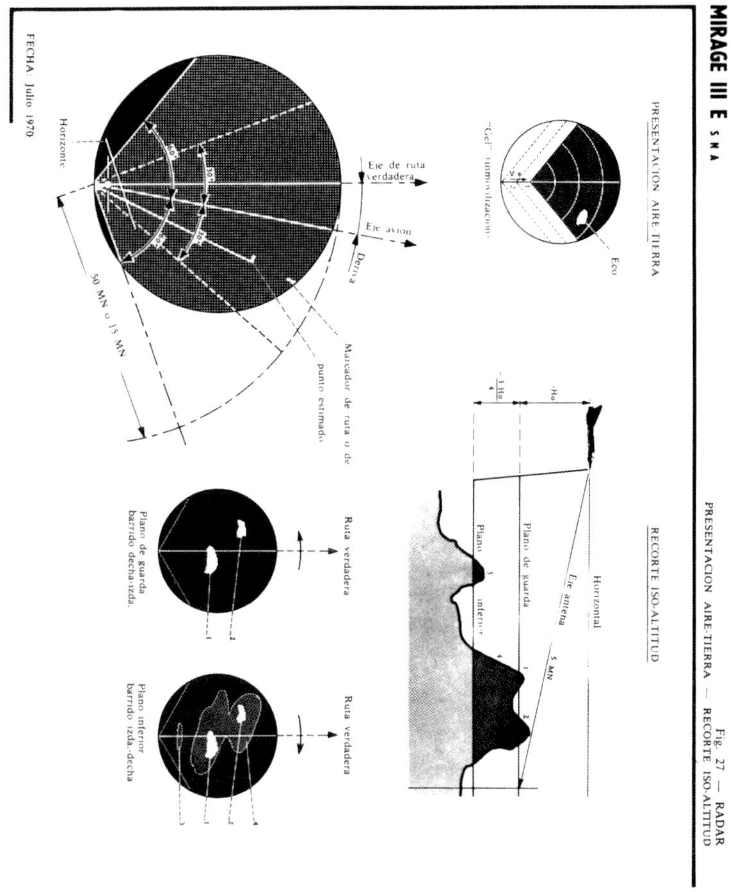

Cyrano IIbis radar blind penetration mode.

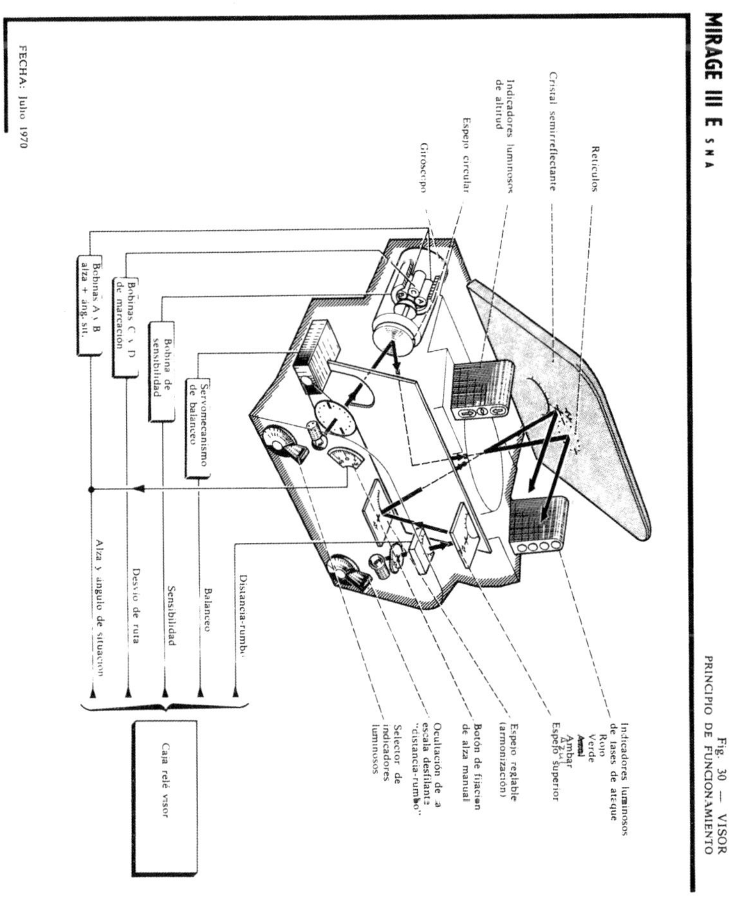

Thomson-CSF 97 electro-mechanical head-up display.

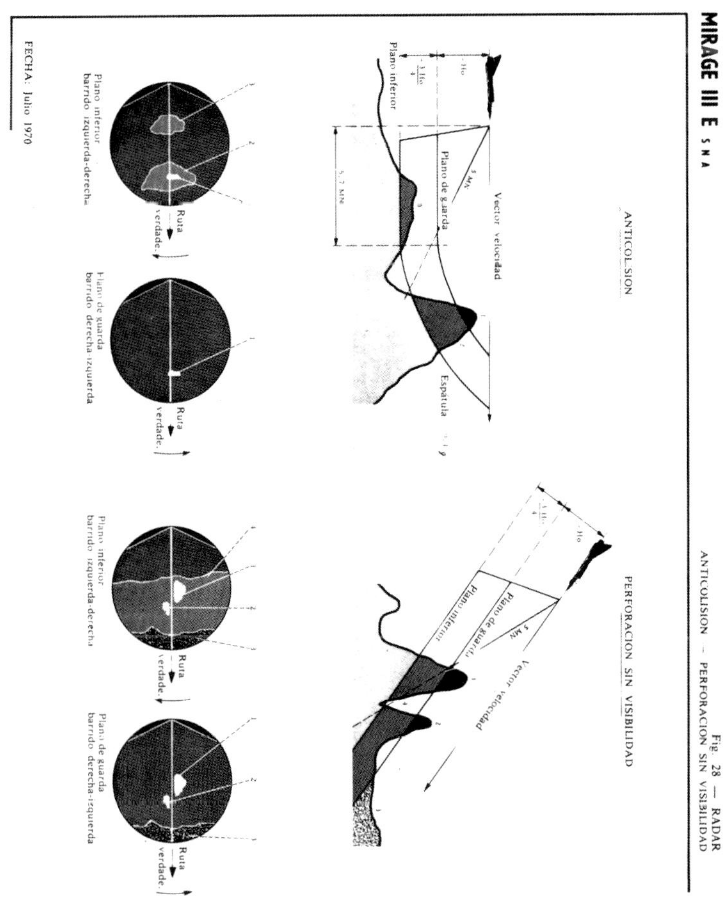

Cyrano IIbis radar air-to-ground sweep.

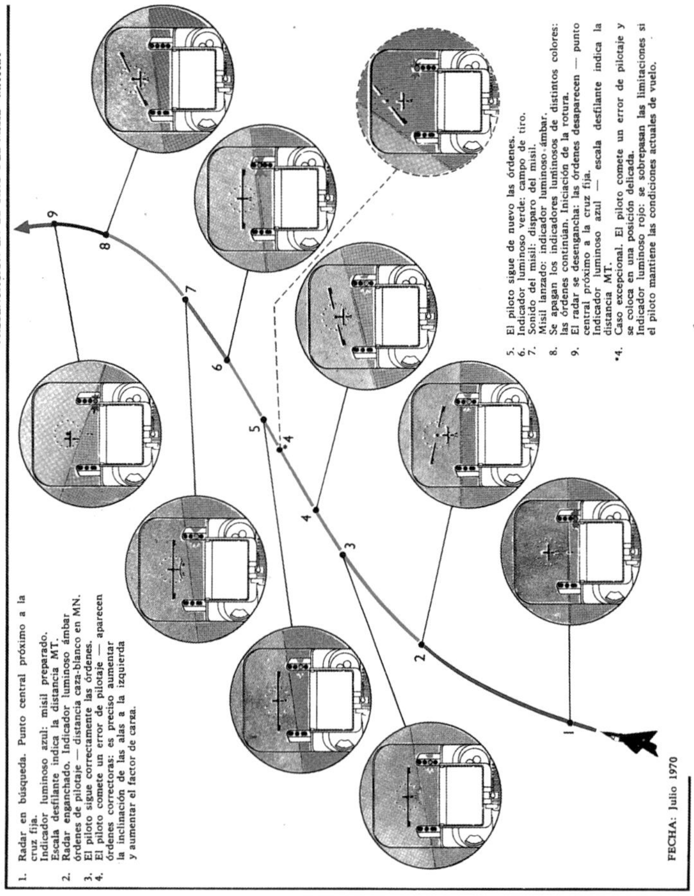

HUD presentation for Matra R530E AAM firing.

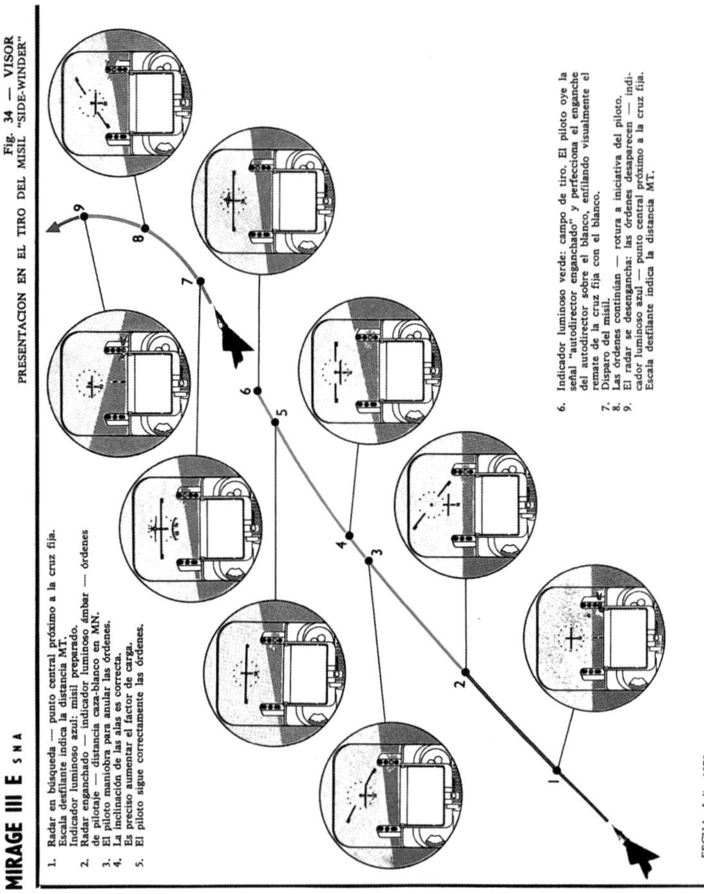

HUD presentation for Sidewinder AAM firing.

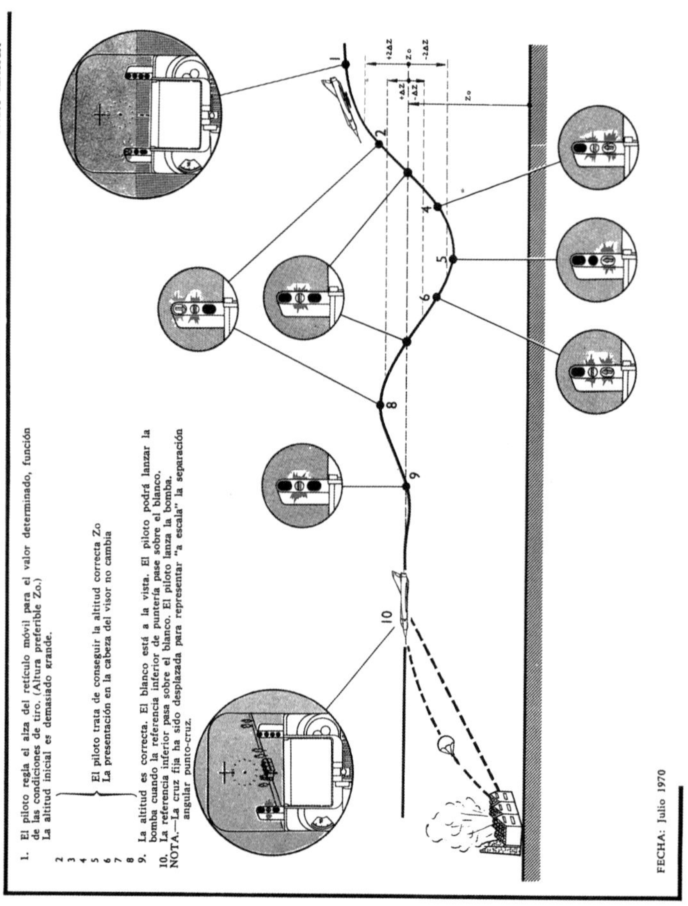

HUD presentation for level bombing.

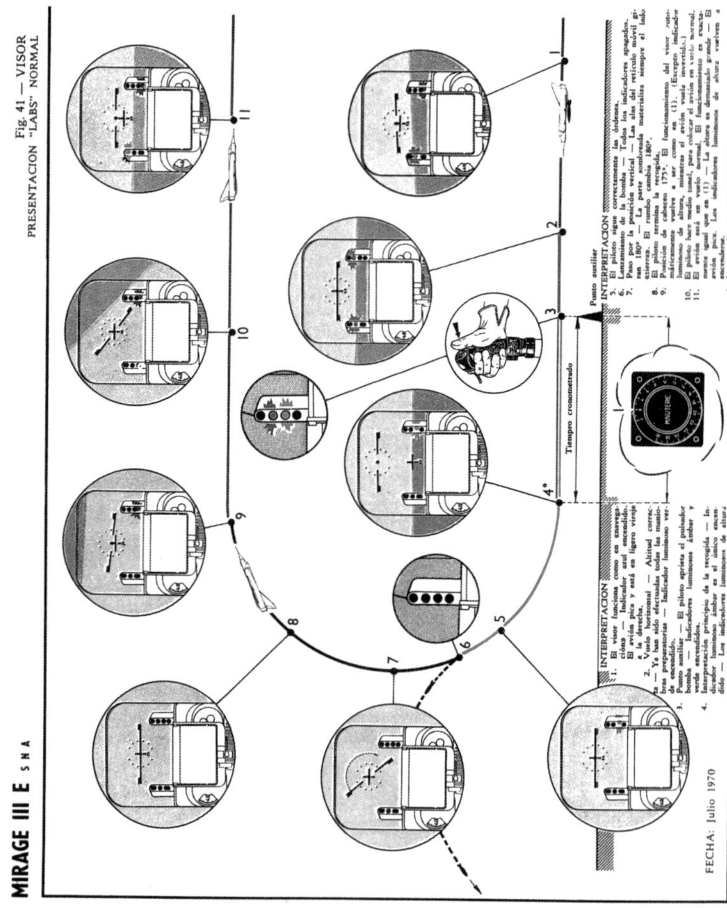

HUD presentation for LABS bombing.

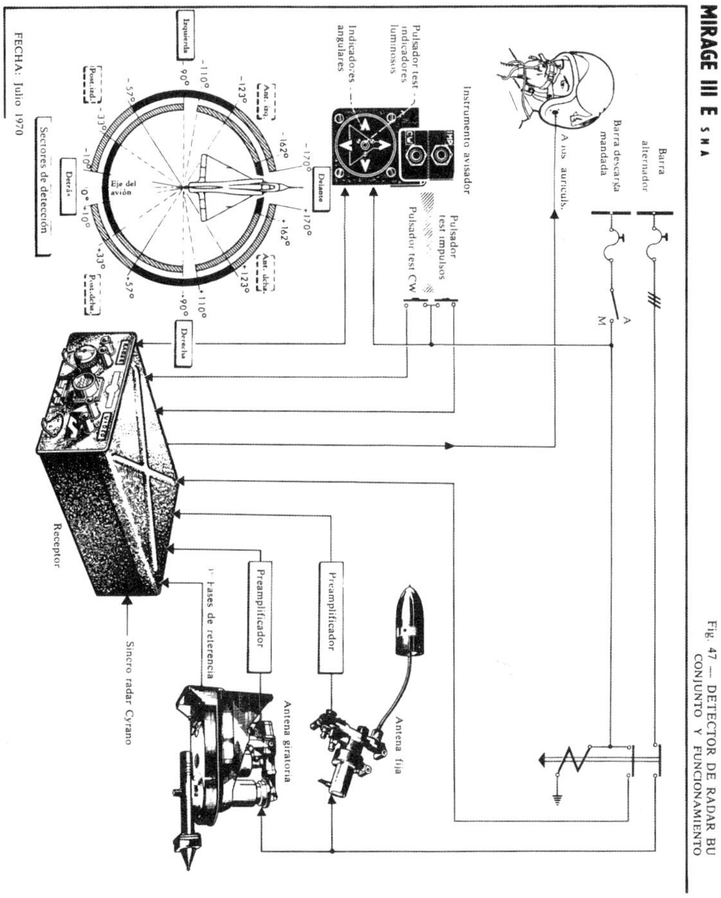

Description of the BU radar warning receiver and its coverage and associated equipment.

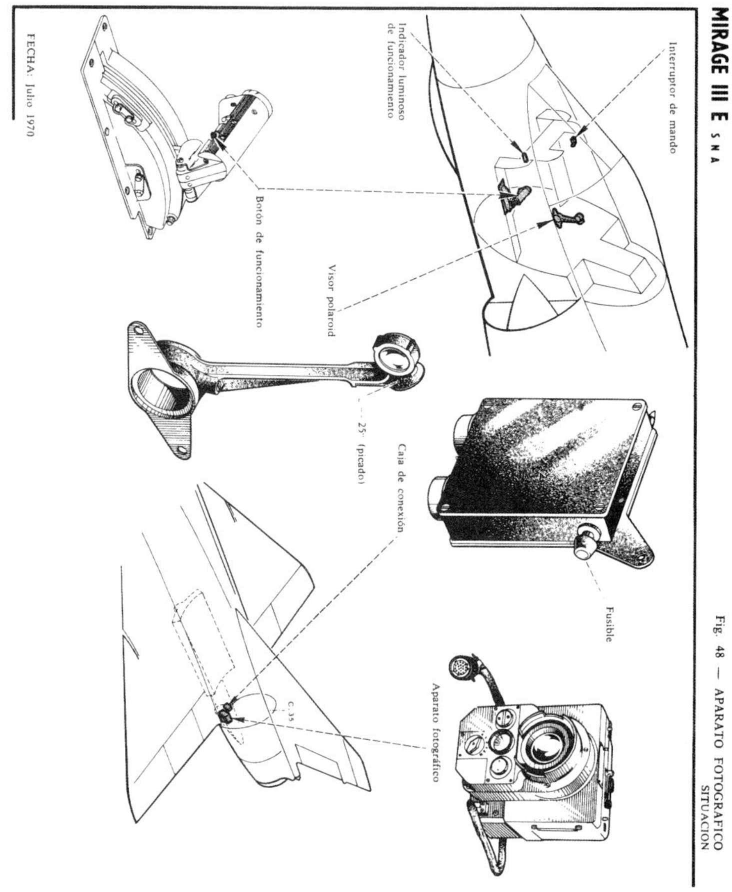

Omera 60 recce camera and associated equipment including cockpit sight.

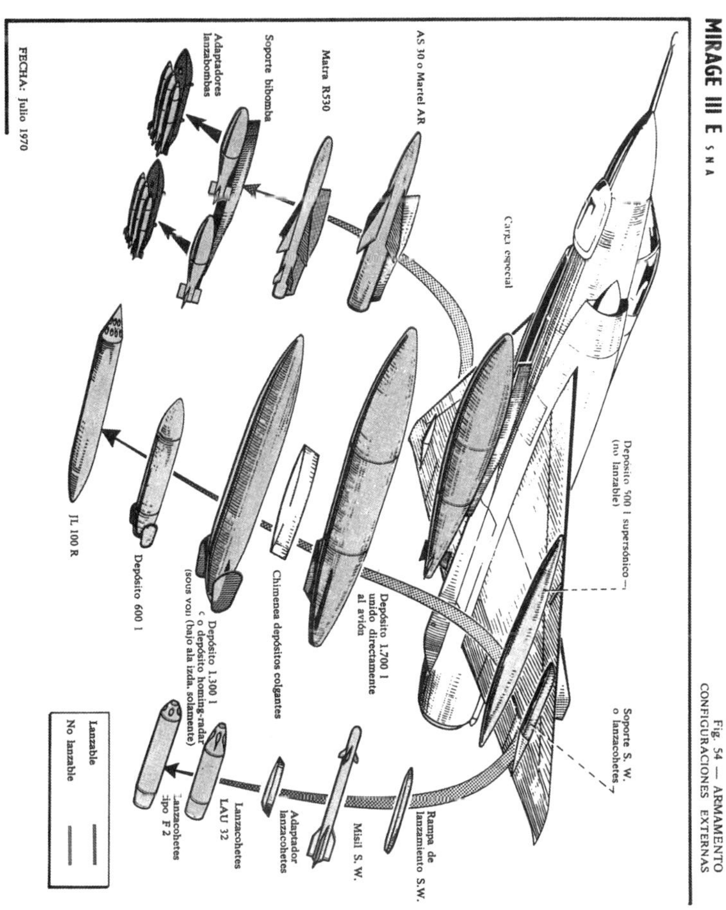

External configurations indicating the stores that can be dropped.

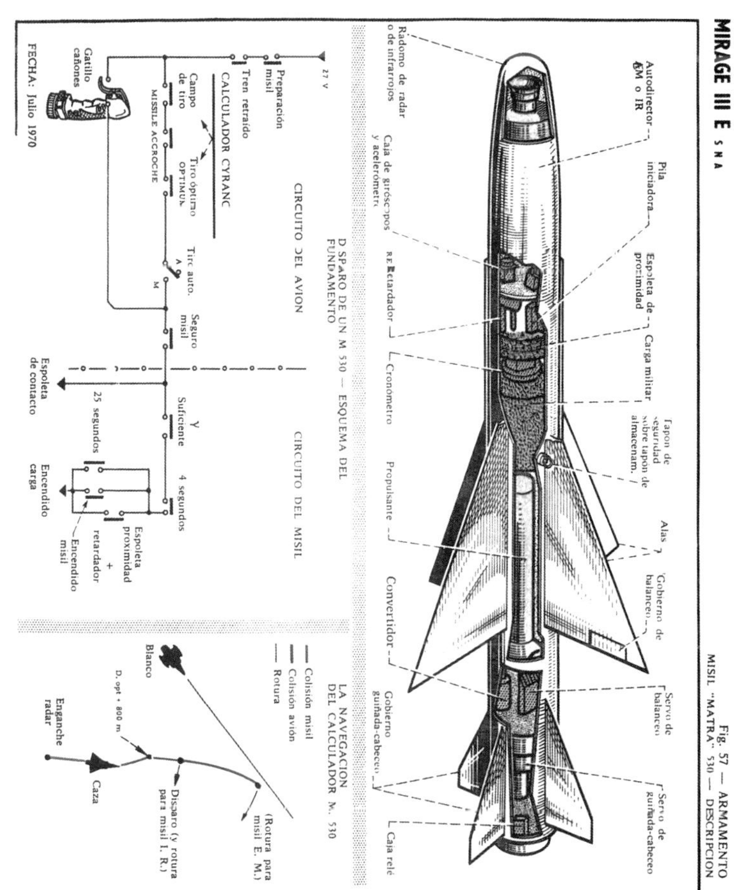

Matra R530E AAM which can be fitted with IR and Semi-active radar seeking heads.

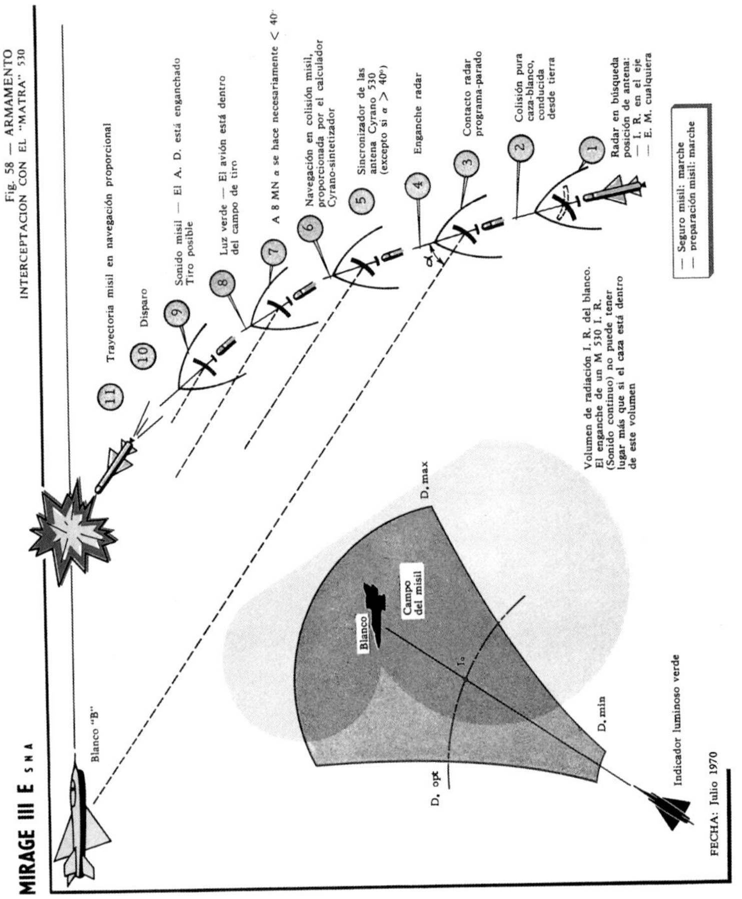

Classic intercept profile with the MATRA R530E.

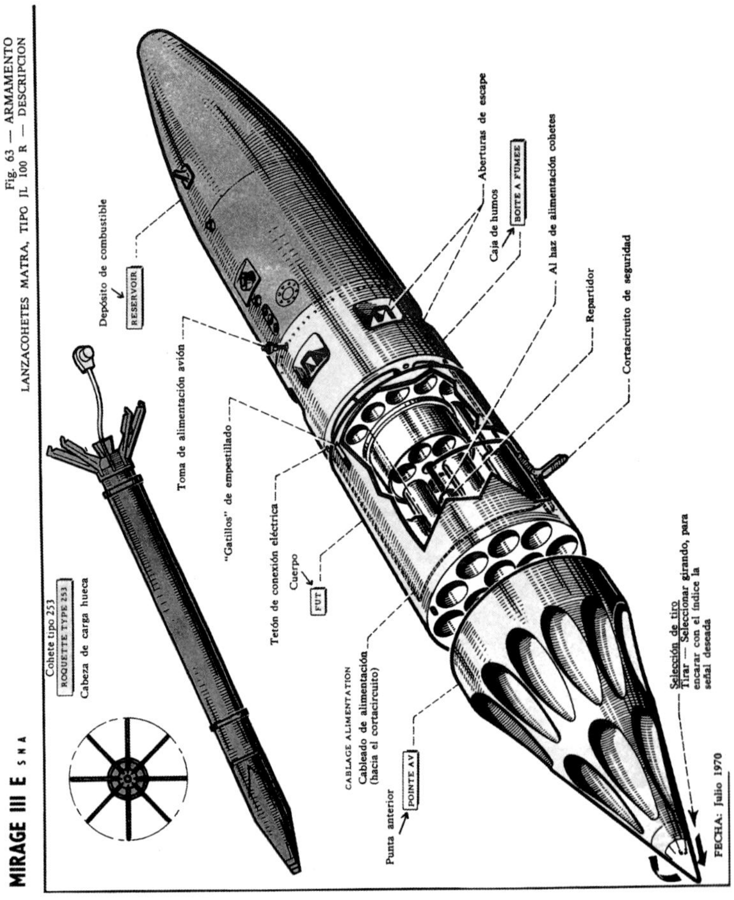

MATRA JL100R rocket pod/fuel tank.

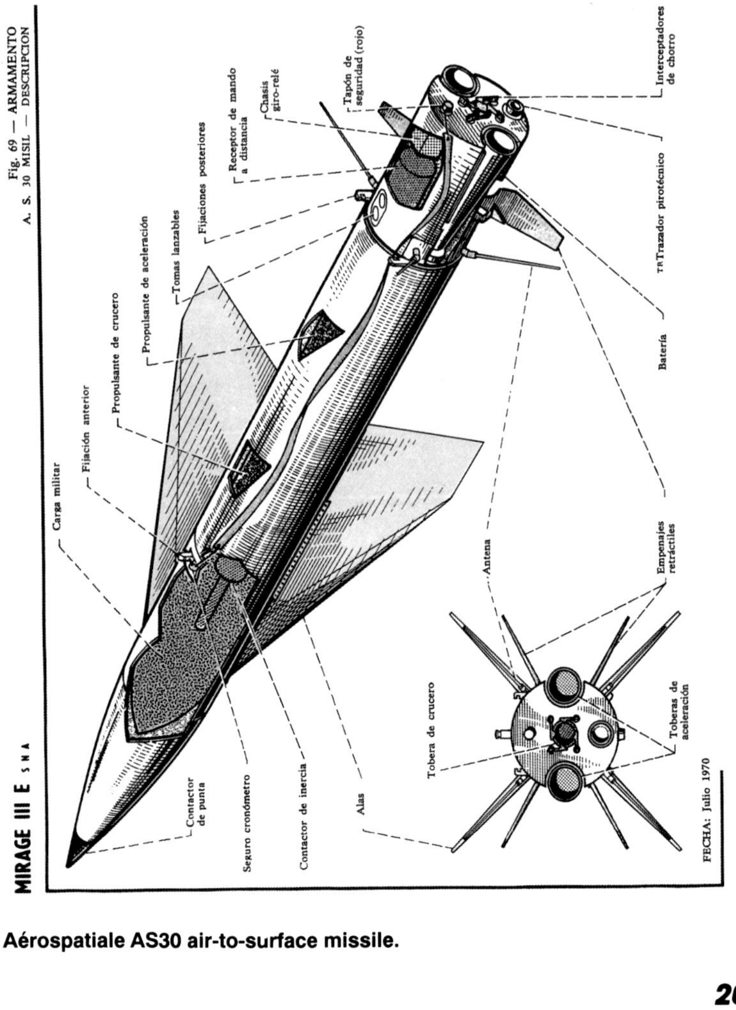

Aérospatiale AS30 air-to-surface missile.